Warship 2022 Conference

Designing Warships for the Future

Bristol, United Kingdom
8 - 9 June 2022

ISBN: 978-1-7138-8309-8

Printed from e-media with permission by:

Curran Associates, Inc.
57 Morehouse Lane
Red Hook, NY 12571

Some format issues inherent in the e-media version may also appear in this print version.

The Royal Institution of Naval Architects
8-9 Northumberland Street
London, WC2N 5DA
United Kingdom

Phone: 020 7235 4622
Fax: 020 7259 5912

publications@rina.org.uk

Additional copies of this publication are available from:

Curran Associates, Inc.
57 Morehouse Lane
Red Hook, NY 12571 USA
Phone: 845-758-0400
Fax: 845-758-2633
Email: curran@proceedings.com
Web: www.proceedings.com

160 YEARS OF HISTORY,
KNOWLEDGE AND EXPERIENCE

SALUM ET CARINAE PIGNORA VITAE

THE ROYAL INSTITUTION
OF NAVAL ARCHITECTS

WARSHIP 2022 CONFERENCE

Designing Warships for the Future

8th-9th June 2022
M Shed, Bristol

Sponsored by:

BMT

THE ROYAL INSTITUTION
OF NAVAL ARCHITECTS

INTERNATIONAL CONFERENCE

WARSHIP 2022

Designing Warships for the Future

8th -9th of June 2022, M Shed
Bristol, UK

Papers published in these conference Proceedings are peer reviewed.

THE ROYAL INSTITUTION OF NAVAL ARCHITECTS
8-9 Northumberland Street
London WC2N 5DA

Telephone: 020 7235 4622
Fax: 020 7259 5912

ISSN No: 2752-6569

CONTENTS

THE VIABILITY OF LOW-CARBON FUELS & GREEN TECHNOLOGIES FOR THE FRONT-LINE NAVAL VESSEL 1
S Newman and T Beard, BMT, UK

EXAMINING THE IMPACT OF FUTURE ALTERNATIVE FUELS ON NAVAL VESSELS 15
R J Pawling, UCL Department of Mechanical Engineering, UK

AN EVALUATION OF A COATING SYSTEM THAT DELIVERS TOP SPEEDS, SUPERIOR MANOEUVRABILITY AND LONG-RANGE CAPABILITIES FOR THE SUSTAINABLE FUTURE OF WARSHIPS 31
M N van Ruiten, Subsea Industries, The Netherlands

FULL AND SEMI AUTOMATED EARLY-STAGE SHIP LAYOUT GENERATION 41
N Hifi and M Courts, BAE Systems Naval Ships, UK

DESIGN FOR AN ALL-ELECTRIC DESIGN FOR SUPPORT IN INITIAL DESIGN USING THE NETWORK BLOCK APPROACH: A SUBMARINE EXAMPLE 51
M H Mukti, R J Pawling, and D J Andrews, University College London, UK

PRACTICAL SHIP DESIGN CONSIDERATIONS FOR MISSION MODULARITY 69
A C Kimber and L Griffiths, BMT Defence and Security UK

OBSOLESCENCE MANAGEMENT TO ENABLE SUPPORTABILITY FOR THE AUTONOMOUS WARSHIP 81
T Flint, BMT Defence and Security Ltd, UK

SURVIVABILITY, ADAPTABILITY AND OFFBOARD SYSTEMS AUTONOMOUS AND OFFBOARD SYSTEMS AND THEIR IMPACT ON WARSHIP AND TASK GROUP SURVIVABILITY 89
D Manley FRINA, RCNC, University College London, UK

COMPREHENSIVE NUMERICAL HYDRODYNAMIC CAMPAIGN TO EVALUATE FRIGATE CONCEPT HULLFORM 97
F Gamboa, BAE Systems, UK

THE DESIGN AND HYDRODYNAMIC ASSESSMENT OF A SUBMARINE CONCEPT WITH OFF-AXIS RESOLVING THE IMPACT OF SHIP'S DESIGN ON DECK AND WELL-DECK LAUNCH AND RECOVERY OPERATIONS BY APPLYING OF QUIESCENT PERIOD PREDICTION (QPP) 111
B Ferrier, Hoffman Engineering, USA
R J Taylor, Royal Navy, UK
M R Belmont, J Christmas, University of Exeter, UK
L Fedrick, MOD, UK

THE ZOOM YEARS: UCL MSC STUDENT WARSHIP DESIGNS THROUGH THE PANDEMIC 133

R J Pawling, UCL Department of Mechanical Engineering, UK

THE VIABILITY OF LOW-CARBON FUELS & GREEN TECHNOLOGIES FOR THE FRONT-LINE NAVAL VESSEL

S Newman and **T Beard,** BMT, UK

SUMMARY

With the global drive to decarbonise, this paper will explore what evolving, largely commercial shipping-led 'green' solutions can realistically be applied to a front-line naval vessel, and what operational compromises must be made to fulfil green ambitions. It will conclude by investigating what a reduced-emission frigate may look like by examining ship design guidelines for methanol fuel.

1. INTRODUCTION

The Net-Zero (NZ) challenge needs a global effort to achieve its objectives. The energy consumed through naval operations, and so the greenhouse gases emitted, is a significant part of the problem and so improvements are needed. In addition, naval vessels often act as travelling embassies so need to exemplify the values and principles of the country they represent, including their commitments to NZ. The decarbonisation of the commercial marine sector is starting to pick up pace, with numerous fuels and technologies being developed, trialled, and implemented. The front-line naval vessel has a unique set of requirements and constraints however which differ to those for commercial shipping. This means it is not possible to just fast-follow the commercial sector- there needs to be tailored solutions that work for defence, ideally taken from, or accelerated by, commercial innovation.

2. NETR-ZERO COMMITMENTS

In 2015, 196 nations signed the Paris Agreement at the United Nations Framework Convention on Climate Change. It committed nearly every nation, including all major carbon-emitting countries, to keeping the global temperature rise this century to well below 2°C rise above pre-industrial levels to limit the devastating effects of climate change. The latest report by the Intergovernmental Panel on Climate Change (IPCC) confirms that this is no longer a challenge for tomorrow- we are already seeing the impacts today. The UK was the first major economy to create a legally binding target to bring greenhouse gas (GHG) emissions to NZ by 2050. In April 2021, through its sixth Carbon Budget, the UK Government set in law the world's most ambitious climate change target, cutting emissions by 78% by 2035 compared to 1990 levels. This would bring the UK more than threequarters of the way to net zero by 2050 [1].

2.1 MARITIME NZ VISION

Commercial shipping is currently responsible for around 3% of global GHG Carbon Dioxide Equivalent [2] (CO_2e) at approx. 1,000 million tonnes annually, and so in support, the International Maritime Organization's (IMO) overarching long-term climate target is to at least halve emissions from the shipping sector (against a 2008 baseline) by 2050, with a 40% reduction by 2030- this is against a backdrop of continually rising numbers of ships.

From a UK perspective, in September 2021 the UK Chamber of Shipping confirmed it wishes (IMO) to double this target and commit to net-zero emissions by 2050. This is a big ask as a study by Lloyd's Register [3], estimates the IMO target of a 50% cut in absolute GHG is equivalent to a real-world reduction of about 85% in operational CO_2 intensity. In other words, ships will have to reduce their output of carbon dioxide by 85% per nautical mile, to take account of increasing numbers of ships, and more activity over coming years.

In 2022 the UK revised the National Shipbuilding Strategy, [4] stating the ambitions for the shipbuilding sector:

- By 2025 all new vessels for UK waters are designed with zero emission capabilities.
- By 2035 bunkering of zero emission fuels is widely available across the UK.
- By 2050 the UK domestic shipping sector is Net-Zero.

2.2 UK DEFENCE AND NZ

Understanding naval shipping's contribution to the GHG problem is difficult to ascertain exactly, but to give context, over the course of 2017 the US Navy alone released around eight million tonnes of CO_2, equating to 1% of that of worldwide commercial shipping. From a UK Defence perspective, the Ministry of Defence (MoD) enterprise contributes close to two million tonnes [5], or 0.5% of the UK's total GHG emissions.

In March 2021, MoD published its "Climate Change and Sustainability Strategic Approach" [6], their vision on the steps needed for UK defence to address climate change. Whilst not addressing the Navy fleet explicitly, it presents defence-wide timescales for implementation of change, with phasing split into "epochs":

- "Epoch one – setting the foundations - 2021-2025: Defence will deliver a step change in the projects that are already planned; initially, most of these will be opportunities on the estate."
- "Epoch two - minimising and fitting for the future – 2026-2035: Defence will look to reduce emissions significantly using existing and emerging technology, maximise opportunity to enhance the global response to the threat that climate change poses to current and future operational capability and build further resilience into the estate, supply chain and future equipment."
- "Epoch three - harnessing the future - 2036-2050: In the final 15-year period, Defence must be doing everything it can to harness novel technologies which further build resilience and further reduce emissions."

So, whilst there is now the commitment from MoD to support NZ, it is not yet clear how the carbon-reduction budget will be flowed down via policy to Royal Navy (RN) for both legacy and new vessels. What is clear though is that each of the Forces will be held accountable to contributing to GHG reduction, RN included.

3. KEY NAVAL REQUIREMENTS

Navies have the following key requirements which differentiate them from commercial shipping, and so drive a different set of constraints into the viability of some NZ options:

- Fuel Supply & Security.
- Underway Replenishment.
- Interoperability with other Navies.
- Battle Survivability.
- Broad Diversity of Vessel Types & Ages.
- Variability of Routes & Roles.

3.1 FUEL SUPPLY & SECURITY

Blue-water navies need to always maintain global operational-capability with its front-line vessels (Carriers, Frigates, Destroyers, Assault and Replenishment vessels). To do so a guaranteed availability of compatible fuel is needed despite any fuel supply issues, wars or other crises which may be happening at the time. This fuel security is met by being independent as far as possible through fuel-storage. For the RN for example, most fuel is issued from the MoD's UK and international strategically located Oil Fuel Depots (OFDs), but fuel of acceptable specification can also be sourced wider within NATO.

From a NZ perspective, this drives the need for front-line naval vessels to be able to embark low-carbon (LC) fuels when available, whilst still retaining the capability to run on the traditional F-76 grade fuel when unavailable, at least for the short-term. Unavailability of a LC-fuel alternative could be due to supply issues caused by immaturity of the LC fuel supply chain, or competition from other sectors such as commercial marine or aviation. For the RN it may be possible for the MoD to use its estate to manufacture its own LC fuels, (as mooted in its Roadmap for Sustainable Defence Support [7]), so bolstering fuel security further, but whether production on such a relatively small scale would be economical needs further consideration.

3.2 UNDERWAY REPLENISHMENT & INTEROPERABILITY CAPABILITY

Replenishment at Sea (RAS) is a critical evolution which all front-line warships conduct, where liquid and solid stores are passed from one vessel to the other, (or from a helicopter to a receiving vessel), whilst connected, whilst underway. NATO interoperability requirements has driven standardisation and so compatibility between member states' RAS systems and fuels. The RAS evolution is hazardous, involving a coordinated, mostly-on-deck effort for both the receiving and supplying vessel crews leaving them particularly exposed. The use of gaseous or cryogenic, volatile or toxic alternative fuels would drive a change to the RAS equipment fitted to all warships and replenishment vessels. Also, safety-management could become more of a challenge, particularly with regards to leaks, spillages and managing emergency break-away procedures.

3.3 BATTLE SURVIVABILITY

Front-line vessels are designed for the battle environment and so the properties of any adopted LC fuels need careful consideration, particularly in the event of a potential release of explosive or toxic fuels through battle-damage, even if during normal operations the risks can be managed as they are for commercial shipping.

3.4 COMPATIBILITY WITH A DIVERSITY OF VESSEL TYPES & AGES

Taking the RN as an example as an 'Owner/Operator', (to use commercial shipping terms), it has a huge range of differing vessel types to manage, from small craft up to aircraft carriers. As such any measures to reducing the carbon footprint will be varied. Over the timeframe of NZ 2050 the RN's front-line fleet will be made up predominantly of in-service vessels or classes under production or planned as shown in Table 1. (Green text shows new classes). Any NZ-facilitating technologies or fuel-change proposals need to recognise this mostly in-service fleet makeup and therefore be retrofittable.

Class	2020's	2030's	2040's	2050's
T83 Destroyers x ?				
T32 Frigates x 5				
(RFA) Future Solid Support Ship x 3				
Future Auxiliary Oiler				
Multi-Role Support Ships x 6				
QEC Aircraft Carriers x 2				
Albion Class Assault Ships x 2				
T45 Destroyers x 6				
T31 Frigates x 5				
T26 City Class Frigates x 8				
T23 Frigates x 13				
(RFA) Wave Class Oilers x 2				
(RFA) Tide Class Oilers x 4				
(RFA) Bay Class LSD x 3				
(RFA) Fort Rosalie Class Solid Support Ship x2				
(RFA) Fort Victoria Stores/Tanker Ship x 1				
(RFA) Argus Casualty Ship x 1				

Table 1: RN Fleet Make-up 2020-2060

3.5 VARIABILITY OF ROUTES & ROLES

Unlike commercial vessels, naval vessels do not usually follow a fixed route of operation, and their role or initial use-case may change through life due to developing operational requirements or emerging threats. This high-variability of operations, often away from the initial design-point proves a greater challenge for the baseline performance prediction and the design and implementation of certain measures to reduce CO_2e, especially for the efficacy of operational-profile-sensitive Energy Saving Technologies (ESTs). This contrasts with a commercial vessel which often conform to a much more fixed operating regime where ESTs, especially those which enable hull or propulsor hydrodynamic improvements, can be optimised to give maximum gains.

4. PATHWAYS TO NET ZERO

Taking the RN as an example, four parallel pathways have been considered as needing to converge to meet NZ. Only LC Fuels and ESTs will be discussed in depth here:

- LC Fuels.
- Behavioural Changes.
- ESTs.
- GHG Offsetting.

The use of LC fuels will have the biggest role to play, as the lifecycle carbon footprint of some can be very low, and so give a step change. Behavioural changes, such as by steaming slower, or any forced as a result of emerging future warfare trends, have the potential to be important NZ contributors, though perhaps the most difficult to quantify. ESTs have worthwhile gains to offer as most technologies are already mature, although the feasibility of implementation and effectiveness is vessel-dependent, and some can't be retrofitted.
Any delta in achieving NZ will need to be addressed by GHG offsetting measures. Whilst there are many, well documented ways organisations can achieve this, for the UK the MoD's considerable estate (around 1% of total UK land mass) is an invaluable resource for offset measures, be it through green energy generation, forestation, or LC fuel production as examples.

4.1 MARINE LOW CARBON FUELS

This section will present the landscape of the LC fuels suitable for marine use, comprising of two groups:

- 'Blend-in' or 'Drop-in' diesel substitutes (DS)- little or no change to safety, operability, equipment, vessel design or infrastructure.
- Non-Diesel substitutes (NDSs)- bigger implications/changes to safety, operability, equipment, vessel design and infrastructure.

As discussed in section 1.4, blue-water navies such as the RN need global fuel-supply security, either through domestic or NATO OFDs, or any other sources if the normal supply chain is disrupted. In the short to medium term at least, LC fuels won't be globally available, and even when they are there will likely be competition from other, more dominant transport sectors such as commercial shipping and aviation. Furthermore, it is likely that the commercial shipping will not be unified in the fuel used. There are currently four low-GHG energy sources that are likely to dominate, these are methanol, ammonia, hydrogen, and batteries. Naval vessels therefore will likely need to have the flexibility to continue to use F76-specification fuel for some time to come, whilst having the ability to use LC fuels where available.

4.1.1 Diesel Substitutes (DS)

Fuels which approximate to or match the existing key characteristics of F-76 clearly have the least impact on current naval platforms. These synthetic DS fuels can be produced from biological sources either grown specifically or from waste, from plastic, or synthesised from hydrogen.

4.1.1.1 Bio-fuels

Biofuels can be broadly broken into groups, categorised in terms of feedstock. First and second generation offer the most promise for NZ timeframes. (Third and fourth generation biofuels are derived from algae but have production-scalability issues). First-generation biofuels are directly related to an oily biomass feedstock that is generally edible such as corn, rapeseed oil, and soyabeans. Second generation biofuels use feedstocks unsuitable for human consumption, and so have a lower potential to threaten food security as the feedstocks can be grown on marginal land or are waste products. Second generation biofuel feedstocks include used cooking oil, lignocellulosic (woody) biomass feedstocks from grasses, renewable wood supplies or other marginal crops, and wastes from forestry, farming and municipal origins.
First-generation biofuels are problematic in that they can cause Indirect Land Use Change (ILUC), which can offset any potential carbon gains. ILUC occurs when pasture or agricultural land previously destined for food and feed is diverted to biofuel production. Food and feed demand still needs to be satisfied, which may then lead to the extension of agriculture land into areas with high carbon stock such as forests, wetlands and peatlands. Such ILUC therefore releases the CO_2 stored in those high carbon-stock areas, and knock-on effects can include damage the environment by displacing wildlife and food-security issues.
FAME (Fatty Acid Methyl Esters) is the most widespread process for the transformation of fats, oils and greases (FOG) feedstocks into biofuels, by breaking the bonds linking the long chain fatty acids to glycerol, replacing it with methanol

in a process called transesterification. FAME biodiesels have several practical use issues, especially for marine use- they have a short shelf life, are hygroscopic and so can suffer from microbial growth, and cannot be used neat in internal combustion engines (ICE) due to their tendency to form deposits and clog filters. As such they need to be blended for use, with a 10% concentration being a typical limit (B10) without requiring any special measures, although 30% (B30) has been trialled successfully by at least 2 commercial operators, TFG Marine and the Mediterranean Shipping Company.
A more advanced processing technique for extracting fuel from FOGs is by hydro (H_2) treating- HVO and HEFA types (Hydrotreated Vegetable Oils and Hydrotreated Esters and Fatty Acids) which offer several benefits over FAME types. These include reduced NOx emission, better storage stability, and better cold flow properties. These improved properties mean hydrotreated biofuels are truly drop-in with no need to blend, though still have the same ILUC issues.
Second generation biofuel feedstocks are generally processed differently and give up their useful compounds less readily, therefore are more difficult to process into useable fuel. This is particularly true of lignocellulosic feedstocks which require a different method involving gasification using the Fischer-Tropsch (FT) process. Like the hydrotreating of oil & fat feedstocks, FT-biofuels are of a high-quality so making them true drop-in fuels.

4.1.1.2 Fuel from Plastic Waste

The production of fuel from waste plastic, (chemical recycling- by pyrolysis), will also form a part of the global future fuels mix. This idea is attractive as currently only a small proportion of waste plastic is mechanically recycled (into new plastic products), with most going to landfill. The efficiency of the processes involved is low however, developing a continuous process is challenging, and the process can run into problems if the feedstock is contaminated with other materials.

4.1.1.3 E-fuels

Known as an electrofuel or "E-fuel", green hydrogen (produced from renewable energy) is reacted with CO_2, either captured from the air, or from waste (e.g. flue gasses) or biomass sources. E-fuel production is very energy intensive, converting less than half of the energy in the electricity into fuel, and only provides clear climate benefits when produced from zero-carbon renewable electricity- but renewable energy generation capacity is limited. The inefficiency of the E-fuel production process means that its use in applications where battery-powered propulsion is viable (such as road transportation) makes little sense. Aviation is likely to be the largest consumer of E-fuels. The RAF has been trialling sustainable kerosene (Sustainable Aviation Fuel, SAF), both from biomass and E-fuel types in its move towards decarbonisation.

4.1.2 Non-Diesel Substitutes (NDS)

Looking beyond the blend-in and drop-in fossil fuel substitutes, several options exist which are being actively pursued in commercial shipping. There are the gaseous (at ambient conditions) fuels of Hydrogen (H_2), Natural Gas (CH_4), and Ammonia (NH_3), and liquid Methanol (CH_3OH).
Hydrogen can be stored onboard as a cryogenic liquid or pressurised gas. It can also be chemically-bound within a Liquid Organic Hydrogen Carrier (LOHC) or a metal hydride. LOHC and metal hydride technologies however are immature and the volume of the onboard extraction processes of these make them unattractive at present. Natural gas can be pressurised for storage but is generally stored onboard in a cryogenic liquid form. Ammonia is generally stored as a cryogenic liquid but can also be stored as a pressurised gas. Methanol is stored under ambient conditions in its liquid state.
Considering the key naval requirements (Section 3), and discounting H_2 chemically-bound within a carrier due to the reasons above, we can begin to assess the suitability of the other NDS candidates as shown in Table 2.

	Hydrogen, H_2	Liquified Natural Gas (LNG), CH_4	Ammonia, NH_3	Methanol, CH_3OH	F-76
With Tank (Gross) Volumetric Energy Density (MJ/L) [8],[9]	2.7 – 7.9	13.2	11.5	14.2 – 15.1	27.3 – 31.0
General Storage Conditions	Cryogenic (or Pressurised)	Cryogenic	Cryogenic (or Pressurised)	Ambient	Ambient
Flash Point	-253°C	-162°C	-33°C	+12°C	+61.5
Flammability Limits in air (vol%) [10]	4.0 – 75.0	5.3 – 15.0	15.0 – 28.0	7.3 – 36.0	0.7 – 5.0
Minimum Ignition Energy in air (mJ) [8], [11]	0.02	0.29	8.0	0.14	20.0
Explosion Risk	Large flammability range with low ignition energy	Medium flammability range with reasonable ignition energy	Medium flammability range with high ignition energy	Medium flammability range with reasonable ignition energy	Small flammability range with high ignition energy
Toxicity	None	None	Highly toxic to humans and aquatic life	Toxic to humans, but very low-toxicity to aquatic life.	
Ease of Vessel Conversion	New tanks required	New tanks required	New tanks required	Existing tanks suitable	N/A – Current Standard
4-stroke Engine maturity	Currently a fuel blend of 20%. Target dual fuel capability by 2030 with aim for 100% H_2.	Available	Development is secondary to methanol	Conversion packages and DF in development.	N/A – Current Standard
RAS-potential	Pressurised or Cryogenic	Cryogenic	Cryogenic, highly toxic	Ambient-condition transfer.	N/A – Current Standard
Availability (Green Variants)	E-fuel requires renewable electricity and water	N/A	E-fuel requires renewable electricity, water, and air	E-fuel requires renewable electricity, water, and CO_2	N/A – Current Standard
Emissions	Potentially high levels of NO_X	Methane slip. Not truly LC as burning contributes carbon	Ammonia slip. Potentially high levels of NO_X	Low	Current Standard

Table 2- NDS Fuel Suitability for Naval Vessels

It is evident from Table 2 that methanol has the greatest in-tank energy density (albeit only half of that of F-76), although ammonia and LNG have similar energy densities whilst hydrogen is significantly lower. It should be noted that LNG is not a 'green' alternative for the long-term goal of NZ, as it still releases CO_2, and more significantly a percentage of methane (methane slip) which has a far greater global warming affect. The other options presented do have the ability to be produced from renewable power. Hydrogen and ammonia produce no COx emissions when used separately, however

there is the potential for them both to produce significant levels of NOx if burnt. LNG and methanol will produce the same emissions as F-76, however on a greatly reduced scale.
All the NDSs presented have low flashpoints which would require any vessel using one to comply with the IMO IGF code. The battle survivability/explosion risk for each fuel is determined based on various parameters, with two key parameters being flammability limit and minimum ignition energy. The minimum ignition energy has greater importance than the flammability limits, since in the event of a missile strike, the higher ignition energy will be more prevalent than the flammable range. The greater the minimum ignition energy then the lower the risk of a deflagration or explosion. There is also the toxic nature of the NDSs to consider, both hydrogen and LNG have limited toxicity to humans or aquatic life. Whilst methanol is toxic to humans, and ammonia is highly toxic to both humans and aquatic life.
There is also the possibility of rapid phase transition for liquid hydrogen, LNG and ammonia. This phenomenon occurs when cryogenic liquid contacts water, the resultant physical explosion could cause significant damage to a vessel. It should be noted that this has been seen on numerous occasions with LNG [12] but not many other fluids, however there is a possibility of this occurring. This is viewed as a significant secondary concern, alongside fire, in the event of a weapon attack.
The integration of the NDSs into naval platforms is of key concern, all of them except for methanol would require custom storage tanks different to that currently used. This is due to the cryogenic nature, or storage pressure for gaseous hydrogen. For cryogenic storage there is also a requirement to maintain the storage temperature, using additional equipment, which can then lead to additional power demands. The technical readiness of engines is also important when determining a potential fuel. 4-stroke LNG engines are available and in use now, whilst methanol is still in development. Ammonia and hydrogen are also in development but at a far earlier stage of the process.
The last aspect which is fundamental to naval operations is the ability to RAS fuel. There is currently no method to dynamically RAS cryogenic or high-pressure fuel. Therefore, significant investment would be required to develop the capability to dynamically RAS hydrogen, ammonia, and LNG. However, because methanol is a liquid at ambient conditions, it is easier to develop a system to dynamically RAS. This is not without problems, because of the low flashpoint and ignition energy a different procedure for emergency breakaway will be required.
It is therefore suggested that for naval vessels methanol is the best option when compared against the other NDSs presented. It can be produced from renewable sources, has the greatest in-tank energy density, fewer concerns for integration and engines are in development. It also has fewer safety implications, when considering survivability from a weapon attack, especially no risk of a rapid phase transition. In addition, it has very low-toxicity to marine life in the event of a spill.
The production of truly green methanol is more energy-intensive than that of hydrogen or even green ammonia, as carbon-capture is required to synthesise it, with direct air carbon-capture technologies being currently inefficient. Carbon may of course be obtained from existing combustion & industrial processes, although if these are not co-located with the methanol production facility then the CO_2 transportation footprint needs to be considered.
In commercial shipping, methanol powered vessels have some pedigree, with the first being Stena Germanica which in 2015 became the world's first RoPax vessel converted to run on both diesel and methanol. With shipping giants Stena and Maersk now building methanol/diesel Dual Fuel (DF) vessels as part of their vessel fleet, others are following suit. As of September 2021, there were 25 methanol-fuelled ships in operation or on order [13].

4.2 ENERGY SAVING TECHNOLOGIES

Below is a summary of the ESTs judged to be most viable for naval applications.

4.2.1 LED lighting

LED lights are considerably more efficient than incandescent or fluorescent alternatives. Not only is the energy required to produce the light lower, but with less heating effect, HVAC cooling loads will also decrease.

4.2.2 Hull Antifouling

Modern antifouling paint systems such as International Paint "Intersleek 900" are already in service with the RN. It is the expectation that future classes will have antifouling coatings applied as standard practice.

4.2.3 Hull Air Lubrication

Air bubbles are introduced to the underside of the hull to reduce friction, with the air provided by a network of pipes and nozzles fed by compressors. The entrapment, distribution & size of bubbles are key to its effectiveness. However V-shaped hulls such as those of the faster naval vessels may not see the same benefits as the slower, flatter-bottomed vessels such as aircraft carriers and replenishment ships, due to the difficulty in maintaining the bubble film. Hull air lubrication will increase underwater radiated noise so increasing a vessels signature above acceptable levels for some naval applications.

4.2.4 Latest Generation ICEs

The use of the latest-generation of engine technology will generally be normal business for new-builds unless a commonality of support solutions drives legacy ICEs to be specified. Retrofit of engines is possible, notwithstanding the spatial and access constraints, T23 PGMU project & T45 PIP projects being current RN examples.

4.2.5 Variable Speed Generation & Direct Current (DC) Bus

By allowing an ICE to run at an optimum speed for its load, rather than being constrained to a fixed speed, (in order for the coupled alternator to output the required shipboard generation frequency), a fuel saving can be realised. By then rectifying the variable-frequency Alternating Current (AC) and feeding the high-power motor driven consumers by a DC bus system (via their own DC-AC power converters) power conversion stages are reduced so resulting in a net energy saving. Such an architecture lends itself to vessels which have a large motor load (either for propulsion or auxiliary systems) when compared to the overall ship electrical demand. A conventional AC 'grid' is then provided via DC-AC conversion for other shipboard consumers.

4.2.6 Energy Storage System (ESS) for Genset Optimisation

ESS using batteries, supercapacitors or flywheels can enable a vessel to operate more efficiently, with two examples being:

- By smoothing power transients which would otherwise be taken up by ramping power on the Diesel Generators (DGs) up and down. For example, a replenishment vessel fitted with an electric RAS system will see oscillating loads on the winch motors due to the relative movement between the two vessels.
- By allowing a ship to run on only one DG for occasions when normally two DGs would be required to be online in order to provide spinning reserve, the ESS provides the reserve instead.

4.2.7 Variable Speed Drives

The use of variable speed drives for pumps offers worthwhile savings, as the motor speed is optimised for the demand as opposed to running the pump-motors flat out at all demand conditions. There are other benefits too, in terms of pump life and reduced fluid noise therefore lower noise signature and improved habitability.

4.2.8 Waste Heat recovery

The inefficiency of prime movers such as DGs leads to a large amount of wasted energy, taken up by the cooling circuits and lost to the exhaust gas. By incorporating Organic Rankine Cycle (ORC) waste heat recovery, or by utilising an Absorption Chiller Plant (ACP), a proportion of this wasted heat energy can be recovered and converted to electrical energy in the case of ORC, or chilled water in the case of ACP. Such systems require additional voluminous equipment, especially if exhaust economisers are installed for ORC, so limiting retrofit & new-build opportunities.

4.2.9 Hull, Propeller & Rudder Features

Hydrodynamic features such as bulbous bows, stern flaps & wedges, propeller boss cap fins and twisted rudders can all potentially offer a reduction to the vessels drag or increase in propulsive efficiency and so a reduction in required powering. Such hydrodynamic features tend to be sensitive to the operating profile however, often having a sweet spot in terms of vessel speed, trim and sea conditions. As stated previously naval vessel rarely conform to regular routes at specific speeds and so gains from these features will often be not as great as those realised for a commercial ship, so these need investigating on a case-by-case basis.

4.2.10 Fuel Cells

Fuel Cells (FCs) are not considered as a viable primary source of front-line naval vessels' power in the timescales of NZ. FCs currently neither deliver the energy density nor fast-enough response time (without being buffered with an ESS, further increasing total system size and weight) required for primary propulsion power- ICEs or gas turbines are the only viable technologies. For a hydrogen or methanol-fuelled vessel (where the fuel may be used directly in the cell without voluminous reforming equipment) they may be viable for ship service power, although the argument for their use is weak where the vessel is fitted with LC-fuelled ICE already, as the ship service power could be more effectively produced using ICE DGs.

4.2.11 Solar Panels

Although photovoltaic (PV) technology is progressing at pace with leaps in efficiency and the advent of flexible and transparent technologies over recent years, their economic application for a front-line warship is likely to be limited. The surface-area required to meet the ships' power requirement, even if just for ship's service load not propulsion, would be considerable, for an application where topside space is generally at a premium. Also to achieve the optimum output from a PV panel it needs to be angled normal to the sun, in doing so the ship's radar cross section may be increased. In addition, any PV solution would need to be buffered by an ESS, would require cleaning to maintain efficiency, and would only achieve peak output for a relatively small percentage of the time.

4.2.12 Wind Devices: Sails, Kites and Flettner Rotors

The variability of routes & roles, coupled with the requirements to minimise RCS, make wind-devices particularly challenging for naval vessels.

4.2.13 Carbon Capture & Storage (CC&S)

Onboard CC&S systems are likely to be too voluminous for most naval platforms, and the subsequent transportation, usage, disposal or storage of the captured CO_2 itself will have a GHG footprint to consider.

5 CLASS GUIDELINES & THE METHANOL-FUELLED FRIGATE CONCEPT

5.1 INTRODUCTION

Following on from the identification of methanol as being an attractive NDS LC fuel for a naval vessel in Section 4.1.2, this final section of the paper will look at what a DF F-76/methanol frigate might look like, and what design, safety & operability measures would be required to realise it.

5.2 SUMMARY OF METHANOL DESIGN GUIDANCE & NAVAL CONTEXT

Currently there are no specific naval vessel Classification Society rules for the use of methanol as a fuel, although rulesets do have sections on Hazardous Area requirements which cover the ventilation, detection, and installed-equipment requirements for spaces which are, or have the potential to be explosive atmospheres. The American Bureau of Shipping (ABS) guidelines for Methanol and Ethanol Fuelled Vessels [14] provides classification criteria for the arrangements, construction, installation and survey of machinery, equipment and systems for vessels operating with methanol or ethanol as fuel. The guidelines do not cover any uniquely naval (military) aspects such as tolerance to battle damage, or RAS-specific requirements (Although bunkering is covered which addresses some RAS aspects).
What will follow is a summary of the key ABS methanol guidelines, with commentary on the naval vessel design implications and relevant interpretation. Where direct quotations from the guidelines are used, these will be presented in quotation marks.

5.2.1 Power & Propulsion Fuelling Redundancy:

- "The propulsion and auxiliary arrangements and fuel supply systems are to be arranged so that in the case of emergency shutdown of the fuel supply the propulsion and manoeuvring capability, together with power for essential services, can be maintained."

DF ICEs are considered to meet this redundancy of fuel supply by their inherent design- ICEs modified for methanol fuelling require a pilot fuel anyway for effective combustion, and so diesel to DF conversions are currently the focus of the engine OEMs. This suits a naval application as the DF approach gives the flexibility of fuels needed to preserve continuity of operations globally.

5.2.2 Guidelines related to Battle Survivability:

Whilst not defining battle survivability requirements, ABS defines the "Limitation of Explosion Consequences", with the explosion in any space containing potential sources of release or ignition not permitted to:

- "Cause damage to or disrupt the proper functioning of equipment/systems located in any space other than that in which the incident occurs."
- "Disrupt the proper functioning of SOLAS equipment or escape routes."

In addition, the guidelines stipulate measures to prevent release following collision or grounding which would inherently give some battle damage tolerance:

- "The fuel tank(s) should be located in such a way that the probability of the being damaged following a collision or grounding is reduced to a minimum."
- "Integral fuel tanks should be surrounded by protective cofferdams, except on those surfaces bound by shell plating below the lowest possible waterline, other fuel tanks containing methanol, or fuel preparation space."
- "Fuel pipes should not be located less than 800 mm from the ship's side."

For a naval application, the expectation is that in order to gain Classification society approval a survivability assessment would be carried out with any specific protection requirements (e.g. ballistic) or other measures derived from that. Naval survivability analysis tools however currently do not have damage algorithms for methanol equipment & systems, so these would need to be developed.

5.2.3 Integral Fuel Tanks, and Piping Arrangements:

- "Integral tanks containing fuel should not be located within accommodation spaces or machinery spaces of category A".
- "For fire integrity, the fuel tank boundaries should be separated from the machinery spaces of category A and other rooms with high fire risks by a cofferdam of at least 600 mm, with insulation of not less than A-60 class". (Cofferdam: a structural space surrounding a fuel tank which provides an added layer of gas and liquid tightness protection against external fire, and toxic and flammable vapours between the fuel tank and other areas of the ship.)
- "Fuel piping that passes through enclosed spaces in the ship should be enclosed in a pipe or duct that is gas and liquid tight towards the surrounding spaces." (i.e., double-walled piping)

And so safety against fuel-system leaks is primarily accomplished through double-walling of systems, either by double walled pipes or cofferdams. For conversions, methanol can be stored in re-purposed existing tanks, though cofferdams may need to be installed depending on their adjacency to certain spaces.

5.2.4 Ventilation & Access:

- "The annular space between inner and outer pipe should have mechanical ventilation of under-pressure type with a capacity of minimum 30 air changes per hour and be ventilated to open air".
- "Any ducting used for the ventilation of hazardous spaces should be separate from that used for the ventilation of non-hazardous spaces".
- Direct access should not be permitted from a non-hazardous area to a hazardous area. Where such openings are necessary for operational reasons, an airlock should be provided.
- "Adequate ventilation is the primary means of dispersal of leaks. Any equipment within the hazardous spaces or gas ventilation systems clearly needs to have the required gas-safe characteristics.." (e.g. sealed or non-sparking) "...to eliminate ignition sources".

Dedicated ventilation of any hazardous or potentially hazardous (in the event of leak) spaces is key to mitigating explosion risk, along with segregation of non-hazardous and hazardous areas.

5.2.5 Nitrogen Inerting:

- "All fuel piping should be arranged for gas freeing and inerting."
- "All fuel tanks should be inerted at all times during normal operation."

In addition to adequate ventilation, the unoccupied space in fuel tanks and other methanol system piping requires inerting using nitrogen. This can either be generated on board or stored in bottles.

5.2.6 Bunkering (& by extension RAS):

- "Suitable protective equipment...is to be provided for protection of crew members engaged in normal bunkering or fuel system maintenance and operations."
- "The connections at the bunkering station should be of dry-disconnect type equipped with additional safety dry break-away coupling/self-sealing quick release."

- "Connections for methanol vapor return lines are to be provided."
- "A ship-shore link (SSL) or an equivalent means for automatic and manual ESD (Emergency Shut Down) communication to the bunkering source should be fitted".

A RAS rig for methanol should not present any practical issues as methanol can be pumped as a liquid at ambient temperatures & pressure. RAS transfer rates would not be constrained by the risk of static build up either, as methanol does not suffer from this. Note the guidelines stipulate that vapour return lines are to be used. This is so that the bunkering vessel recovers the displaced gas from the tank being loaded where it is stored, treated, or disposed of under controlled conditions. If vapour return is deemed mandatory for a naval applications, then parallel methanol supply and vapour return hoses would be required to be passed between the bunkering and receiving vessels.

5.3 METHANOL-FUELLED FRIGATE CONCEPT

BMT's Venator 110 vessel design has been used as the basis, with a concept arrangement of the methanol fuel system presented, repurposing some of the existing F-76 tanks. This gives a reduced net emission capability, whilst maintaining operational flexibility with respect to fuel used.
Both the certification rules and current ICE technology drive the requirement to still carry a quantity of F-76 as pilot/backup fuel. The required proportion of F-76 to methanol onboard will be dictated by factors such as carbon reduction target, required range, and anticipated methanol availability on the routes the vessel is required to perform. Also, there will be a minimum F-76 volume which driven by the pilot-fuel ratio required by the engine. For this concept, where the existing baseline design is adapted with minimum change, the three forward-most tanks are used for methanol with the two aft tanks reserved for F-76, giving approximately a 60%/40% ratio of F-76 to methanol. This approach allows the vessel to comply with existing class power & propulsion survivability and redundancy requirements, as it maintains two segregated, independent F-76 supply systems.

Key Features (See Figure 1):

1. Dual-Fuel (Diesel-Methanol) 4-stroke-engines and generator sets. On-engine methanol systems are double-walled.
2. Aft F-76 Dieso tanks are maintained for pilot fuel & as a backup in case of methanol fuel system issue or safety shutdown.
3. The Fuel Preparation Space (FPS) contains all the fuel conditioning systems required to provide methanol to the engines at the required temperature and pressure. The FPS must be outside the main (Cat-A) machinery spaces. As it is impractical to 'double-wall' such equipment, the whole space is classed as a hazardous area.
4. The methanol tanks are the forward F-76 tanks re-purposed. Zinc coating is applied to give corrosion resistance. The tank void space above the fuel is kept inert with nitrogen to displace the volatile fumes
5. Double-walled piping is used between fuel preparation space and engines, with the annulus void ventilated, allowing all areas downstream of the FPS to be considered non-hazardous.
6. Dedicated ventilation arrangements for overpressure of tanks and FPS. The ventilation systems have strict requirements in terms of flow rate, velocity and entry and exit locations.
7. Cofferdams to separate methanol tanks and non-methanol tanks, and to separate tanks from non-hazardous spaces.

From an operational perspective, the range of the vessel will be compromised with respect to a purely F76-fuelled equivalent design, due to the reduced volumetric energy density (approximately half that of F-76) of the methanol. Whilst it may be possible to design naval platforms with enough methanol capacity to offset some of the range deficit, it is unlikely that the full deficit could be recovered by doubling tank capacity, as the knock-on compromises that would be made to the vessel design would be intolerable.

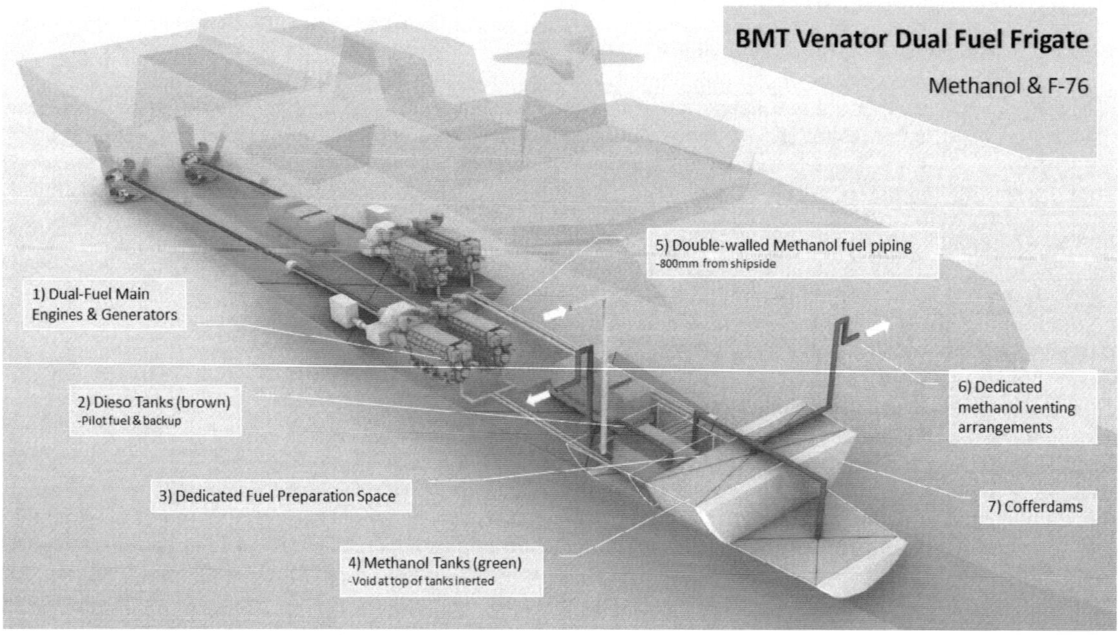

Figure 1- BMT Venator Dual Fuel Frigate Concept

5. CONCLUSIONS

The unique set of requirements & constraints under which Navies operate means LC-solutions for commercial shipping aren't always viable, or effective. However, as commercial shipping is driving the timeline for maritime LC developments at pace, it makes sense that Navy become fast-followers to those solutions which are appropriate.

Platform-level NZ targets will not be met through one measure, with a mix of fuel-saving behavioural changes plus ESTs needing to compliment the far more impactful shift to LC fuels.
To establish the future naval LC fuel, an understanding of future fuel global availability and economics is required, a true picture of LC fuels' 'well-to-wake' emissions gained, and any decision needs to consider interoperability needs. Given that DS fuels offer the minimum vessel and operational impact, focus should be given to securing those. However, if an NDS such as methanol becomes the LC fuel of choice for navies going forward, then there will be operational compromises such as reduced range which will have to be accepted, along with significant adaptation costs both at platform and RAS-infrastructure levels.

The BMT Venator 110 Methanol concept presented demonstrates the key considerations required to integrate methanol capability into a naval vessel, with methanol ship conversations being proven in commercial shipping. Battle survivability requirements are unique to naval vessels, so it is expected that Class-approval for the use of an NDS will be via an extension of the standard survivability assessments required for naval platforms, although additional gas leakage and explosion algorithms may have to be developed for the survivability modelling.

6. REFERENCES

1. UK Government, 'UK enshrines new target in law to slash emissions by 78% by 2035', *Press Release*, 20 April 2021.

2. International Maritime Organisation, 'Fourth IMO Greenhouse Gas Study', 2021.

3. Lloyds Register & University Marine Advisory Services, 'Transition Pathways, 2019.

4. National Shipbuilding Office, 'National Shipbuilding Strategy', 2022.

5. National Audit Office, 'Ministry of Defence Environmental Sustainability Overview', May 2020.

6. Ministry of Defence, 'Climate Change and Sustainability Strategic Approach', March 2021.

7. Ministry of Defence, 'Roadmap for Sustainable Defence Support', August 2020.

8. MariGreen, 'Perspectives for the Use of Hydrogen as Fuel in Inland Shipping', October 2018.

9. Van Hoecke, Laffineur, Campe, Perreault, Verbruggen, Lenaerts, 'Challenges in the use of hydrogen for maritime applications', *Royal Society of Chemistry*, Jan 2021.

10. Lewis, Von Elbe, 'Combustion, Flames and Explosions of Gases', 1987.

11. Verkamp, Hardin, Williams, 'Ammonia combustion properties and performance in gas-turbine burners', *International Symposium on Combustion*, 1967.

12. Melhem, Saraf, Ozog, 'LNG Properties and Hazards', *ioMosaic Corporation*, 2006.

13. DNV, 'Alternative Fuels for Naval Vessels', 2021.

14. ABS, 'Methanol and Ethanol Fuelled Vessels', January 2022.

7. AUTHOR BIOGRAPHIES

Simon Newman is a Principal Engineer at BMT Defence & Security UK Ltd, specialising in marine propulsion system design with an interest in green technologies. Previous experience includes the specification & integration of the MT30 gas-turbine into the Queen Elizabeth Aircraft Carriers.

Dr. Thomas Beard is a Senior Engineer at BMT Defence & Security UK Ltd, specialising in future fuels and survivability analysis. Previous experience includes a doctorate in hydrogen safety and integration of various equipment into surface ships and submarines.

Warship 2022: Designing Warships for the Future
8th-9th of June 2022
M Shed, Bristol, UK

EXAMINING THE IMPACT OF FUTURE ALTERNATIVE FUELS ON NAVAL VESSELS

R J Pawling, UCL Department of Mechanical Engineering, UK

SUMMARY

Emissions regulation aimed at reducing air pollution, particularly carbon dioxide emissions, is driving commercial research into alternative fuels such as ammonia, methanol and hydrogen. Whilst naval ships, being government owned vessels, are exempt from this regulation, the ability to operate on such fuels may become a requirement for future combatants. This paper describes an ongoing study to examine the impact of alternative fuels over a range of naval vessel designs and sizes, using the ZEOLIT early-stage ship tool. The initial stage has examined an OPV, as this is a relatively simple type of vessel, which has already seen the adoption of dual fuelling with LNG in some navies. Three future fuels have been investigated: methanol, ammonia and a Liquid Organic Hydrogen Carrier. The findings so far indicate that, for an OPV, these fuels lead to an increase in displacement, but not one that would render the ship impractically large.

1. INTRODUCTION

Emissions regulation aimed at reducing air pollution, particularly carbon dioxide emissions, is driving commercial research into alternative fuels such as ammonia, methanol and hydrogen. Whilst naval ships, being government owned vessels, are exempt from this regulation, the ability to operate on such fuels may become a requirement for future combatants and dual-fuel arrangements have already entered into operation on patrol vessels, which spend much of their time within ecologically sensitive home waters. These fuels present major challenges for naval use, such as lower energy density, increased toxicity, increased flammability and explosion risk.

Recent UCL MSc ship designs have been set the requirement to operate on future fuels likely to be found for commercial use, the concept being that cost and availability will drive adoption of these energy sources during peacetime, with reserves of marine diesel oil / NATO F-76 retained for use in wartime where its safety advantages are most valuable. These studies indicated that dual-fuel concepts are viable, and they are described in a separate paper [1]. For example ammonia will require large cylindrical tanks for storage as illustrated in Figure 1.

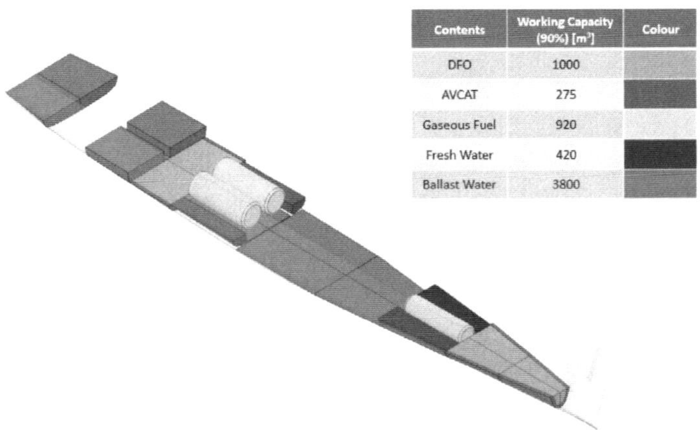

Figure 1: UCL MSc student design for a dual fuel landing ship, showing tankage arrangements with large cylindrical ammonia tanks [1]

Being point designs developed in a teaching context, there was limited opportunity to determine the "delta" in design size and cost for adoption of the dual-fuel approach. Accordingly, a separate ongoing study is underway to examine the impact of alternative fuels over a range of naval vessel designs and sizes, using the ZEOLIT early-stage ship tool first described in [2] and [3]. The objectives of this study are both to contribute to the ongoing discussion regarding future fuels for warships and provide additional information and guidance for UCL MSc students.

2. THE ZEOLIT DESIGN TOOL

ZEOLIT (Zonal Exploration Of Layout and Impacts Tool) is a UCL-developed, Excel-based tool for the early-stage design of ships. It was previously described in [2] and [3], where it was applied to the investigation of the impact of EM railguns on small combatants. ZEOLIT was developed from a thought experiment on whether some types of early-stage design studies might be best carried out with a tool that deliberately limited the level of detail, in contrast with many sophisticated approaches that seek to add more detail. This was in part inspired by the discussion by McKesson regarding simple models in design [4]. ZEOLIT divides the ship into 32 regions; 8 sections along the length of the ship, and four levels (lower hull, upper hull, lower superstructure and upper superstructure). This division is illustrated in Figure 2.

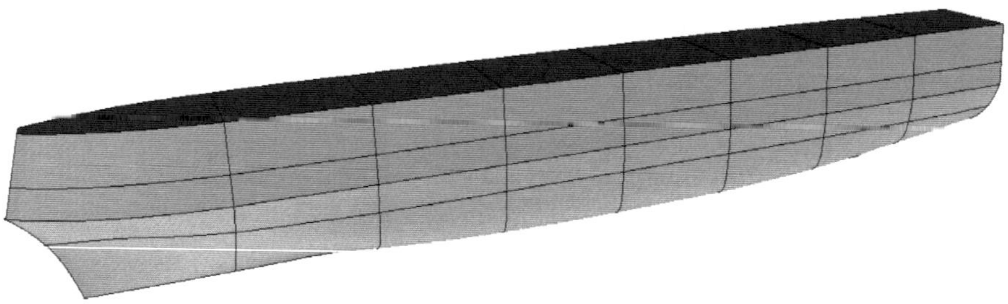

Figure 2: Illustration of the generic envelope with 32 regions

The proportion of each region that is available is defined by a "mask", which for the hullform regions is usually "1" per region, but is lower for superstructure regions. Spaces are allocated to these regions on a simplified profile view, with distributed spaces being assigned fractional amounts per region. Importantly, ZEOLIT does not attempt to iteratively balance the design. Instead, it examines every entry in a database of envelopes, covering a range of displacements and hullform coefficients, retaining only those options where the balance of weight and space are within user-specified limits (usually +/-5%). To check the small ship database of 5832 hull options took just under 5 minutes on a 4GHz 8-core Ryzen 7 PC. Table 1 from Pawling et al [2] summarises the approach to defining the ship envelope and layout in ZEOLIT.

Table 1: Layout modelling in ZEOLIT [2]

Reference Volume

	Stern							Bow	
	H	G	F	E	D	C	B	A	*19393*
2 SS H	799	992	1338	922	1370	816	749	309	*7295*
1 SS L	420	522	704	485	721	430	394	163	*3840*
-2 No 2	446	554	747	515	764	456	418	172	*4072*
-3 H L	123	397	728	712	977	734	432	83	*4186*
19393	*1788*	*2465*	*3517*	*2635*	*3832*	*2435*	*1994*	*727*	

Pre-calculated hull option, showing the possible internal volume for the hull and a notional full-height-full-width-full-length single block superstructure.

Geometry Disposition

	Stern							Bow	
	H	G	F	E	D	C	B	A	*20*
2 SS H	0	0	0	0	0.5	0	0	0	*0.5*
1 SS L	0	0	1	1	1	0.5	0	0	*3.5*
-2 No 2	1	1	1	1	1	1	1	1	*8*
-3 H L	1	1	1	1	1	1	1	1	*8*
20	*2*	*2*	*3*	*3*	*3.5*	*2.5*	*2*	*2*	
Number	21								

"Mask" of the available envelope, which defines how much of the pre-calculated hull is actually used. In this example the hull has a single superstructure block amidships with a small bridge deckhouse (layout zone D-2).

Item Distribution Inspector

	4								
	Stern							Bow	
	H	G	F	E	D	C	B	A	*1*
2 SS H	0	0	0	0	0	0	0	0	*0*
1 SS L	0	0	0	0	0	0	0	0	*0*
-2 No 2	0	0	0	0	0	0	0	0	*0*
-3 H L	0	0.25	0.1	0	0.25	0.4	0	0	*1*
1	*0*	*0.25*	*0.1*	*0*	*0.25*	*0.4*	*0*	*0*	

Inspecting Liquid Fuel Tanks
Number 4

Spaces and equipment are then assigned to layout-zones. Distributed items, such as fuel tanks, can be assigned to multiple zones as fractions. Alternatively, a pseudo-random assignment could be used where the designer specifies the probability of being in a certain layout zone.

Current Volume
FG: move

	Stern							Bow	
	H	G	F	E	D	C	B	A	*2406*
2 SS H	0	0	0	0	187	0	0	0	*187*
1 SS L	0	0	0	0	0	0	0	0	*0*
-2 No 2	0	0	104	27	0	0	0	0	*131*
-3 H L	255	187	355	759	258	274	0	0	*2088*
2406	*255*	*187*	*459*	*786*	*445*	*274*	*0*	*0*	

The current volume required in each layout-zone can be calculated, shown here for the entire "move" function (i.e. machinery, fuel tanks etc.)

Current Total Volume
Update totals

	Stern							Bow	
	H	G	F	E	D	C	B	A	*9280*
2 SS H	0	0	0	0	332	0	0	0	*332*
1 SS L	0	9	575	447	474	163	9	0	*1676*
-2 No 2	448	496	621	428	647	401	273	443	*3758*
-3 H L	275	435	504	954	381	466	303	196	*3514*
9280	*723*	*940*	*1699*	*1830*	*1834*	*1030*	*584*	*639*	

This aggregation can be repeated over all functions and all layout zones to calculate the total volume (and weight) in each zone.

In addition to the overall balance of supply and demand for internal volume and weight/buoyancy, ZEOLIT records the total balance of volume supply and demand for each region. This can then be examined to determine if the simplified layout is acceptable. Results are visualised via scatterplots and statistical metrics, for example the plots of estimated GMts and distribution of hullform proportions shown in Figure 3. ZEOLIT is intended to determine general characteristics of a concept, rather than of a point design.

Input Values			
Circular M	B/T	T/D	flare
6	3	0.4	6
7	3.5	0.45	
8	4	0.5	8
Occurance % Counter			
37	43	3	51
34	32	27	
29	25	70	49

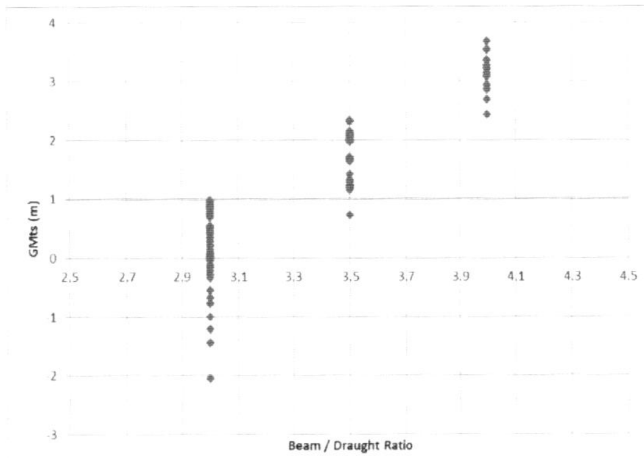

Figure 3: Example ZEOLIT outputs, showing the occurrence of different hullform parameters in the results set (left) and the range of estimated GMts (right)

3. FUTURE FUELS

3.1 THE FUTURE FUEL LANDSCAPE

The current global marine fuel landscape is quite diverse, with five main liquid fuels (Marine Gas Oil, Marine Diesel Oil, Intermediate Fuel Oil, Marine Fuel Oil and Heavy Fuel Oil), LNG gas fuel and a wide range of derivative specifications, such as NATO F-76, a type of MDO. However, these have a common source and generally well-established use cases based on physical properties and cost. The literature contains many surveys of possible future marine fuels by classification agencies [5], academia etc. [6] and these are marked by significant variation in source, physical and financial characteristics and prime mover suitability.

3.2 POSSIBLE FUTURE NAVAL FUELS

The future fuel options for naval ships have been discussed in both general survey terms, e.g. [7] and more specific examples, such as methanol [8] and Gas-To-Liquid [9] concepts. LNG has already seen some use on Ocean Patrol Vessels, such as the Finnish Turva and Norwegian Berentshav class, which have dual-fuel engines capable of using LNG or diesel. Submarines with Air Independent Propulsion (AIP) using fuel cells usually run on hydrogen (e.g. German Type 212) or alcohols (e.g. the ethanol fuelled MESMA system). The key characteristics of some main fuels of interest for the naval sector are summarised in Table 2.

Table 2: Characteristics of some alternative fuels of interest

Fuel	Typical density (te/m^3)	Typical Lower Heating Value (MJ/kg)	Storage
F-76 (reference)	0.85	41	Atmospheric pressure and temperature, arbitrary geometry
Methanol	0.79	20	Atmospheric pressure and temperature, arbitrary geometry, dry air
Ammonia	0.682	18.6	High pressure tank -or- chilled to -33.6deg C, atmospheric pressure
LNG	0.45	49	Cryogenic tank
Liquid Hydrogen	0.07085	120	Cryogenic tank

GTL	0.78	44	Atmospheric pressure and temperature, arbitrary geometry
LOHC	0.984	5.98 *	Atmospheric pressure and temperature, arbitrary geometry

* for benzyltoluene containing 50kg hydrogen per tonne

3.3 LIQUID ORGANIC HYDROGEN CARRIERS (LOHC)

The energy-generating fuel for fuel cells is hydrogen, with fuels such as ammonia or methanol being hydrogen carriers, using the heavier nitrogen or carbon atoms to increase the fuel density. The hydrogen must be liberated for use in the fuel cell by endothermic reactions (dehydrogenation or reforming). High temperature cells such as HT-PEM or Solid Oxide fuel cells (SOFC) are sometimes regarded as being "auto-reforming" as this process takes place internal to the fuel cell module. An alternative concept, however, is the Liquid Organic Hydrogen Carrier, or LOHC.

LOHCs are complex hydrocarbons that can be reversibly de- and re- hydrogenated, as opposed to the simpler molecules such as methanol, which are reduced to hydrogen, water and CO_2. The de-hydrogenated form of the LOHC will generally have broadly similar physical properties and in many cases the chemicals involved are common solvents. The two main problem with LOHCs are the small amount of hydrogen carried by weight, and the large amounts of energy (heat) needed to drive the reaction, although research continues on catalysts that would reduce the latter requirement [10]. LOHCs are generally specified as a pair of chemicals, examples being toluene/methylcyclohexane, dibenzyltoluene/perhydro dibenzyltoluene or benzyltoluene /perhydro benzyltoluene [11]. Methylcyclohexane will be familiar to some readers as it is used as the solvent in Tipp-Ex.

3.4 ZERO CARBON VS. NET ZERO VS. ZERO EMISSIONS

A fuel that contains no carbon, such as hydrogen or ammonia, will produce no CO_2 emissions in use. However, it may still have significant emissions. Hydrogen has a global warming potential of 5.8 times CO_2 over a 100-year period and a hydrogen economy will lead to releases via leaks etc. Additionally, the exhaust from fuel cells can contain unreacted fuel. Combustion processes using atmospheric air will always produce nitrous oxides (NOx), and nitrogen-rich fuels such as ammonia will increase this. This will require improvements to scrubber performance to remain within existing regulations.

An additional consideration for the commercial user is whether regulations permit "net zero", where a carbon-containing fuel such as methanol is acceptable, along with its' CO_2 emissions during use, so long as it is sourced using atmospheric carbon dioxide, e.g. biofuels or electrofuels. Both of these production processes have significant infrastructure requirements, being land or energy intensive. A more significant issue from the commercial side may simply be certification. The international transport of biofuels has already proven controversial as it can result in emissions simply being outsourced [12] and achieving a reliable accreditation process via IMO may be difficult. However, it may be easier for governments or alliances such as NATO to provide accreditation of the "net-zero" status of future fuels operating within a closed logistics system given that F-76 is already subject to stricter quality controls than the various grades of commercial fuels.

4. MACHINERY ASPECTS

4.1 POWER DENSITY

Warships are marked by significantly greater installed power density than commercial vessels, even those with very high hotel loads such as cruise liners, as illustrated in Table 3. This tends towards a preference for lighter propulsion machinery, at the cost of increased fuel consumption. The second major difference is that warships are operated over a wide range of propulsion powers, rather than long periods at a single speed. However, the common perception of commercial vessels as only being operated at a "ideal" single speed is increasingly inaccurate, with "slow steaming" and "flexible steaming" seeing a much wider range of speeds adopted [13]. The more significant difference is perhaps that warships need to rapidly change their power usage, both in terms of speed and hotel load, particularly for modern high-power weapons. Waste heat recovery systems, which are common in commercial vessels, can be difficult to integrate into warships for this reason, as discussed in [14]

Table 3: Example power density in MW (power and propulsion combined) per deep displacement tonne

	MW/te
Destroyers	0.0078
Modern Frigates	0.0052
Cold War Frigates	0.0104
OPVs	0.0086
Carriers	0.0018
Container Ships	0.0004
Cruise Liners	0.0013

4.2 PRIME MOVER OPTIONS

Table 4 provides some representative characteristics of prime mover types of interest to the naval sector, including near-term options such as Proton Exchange Membrane (PEM) and Solid Oxide (SO) fuel cells. It is immediately apparent that fuel cells are heavier and larger than internal combustion engines, but much more efficient. However, their "true" fuel is hydrogen, which must be extracted from a hydrogen carrier (e.g. methanol) by reformation, a high-temperature, endothermic process. High temperature PEM and SOFC are capable of reforming their own fuel and the values in Table 4 include reforming equipment. More detailed comparisons of fuel cell types can be found in the literature, e.g [15]

Table 4: Characteristics of some prime mover types of interest

Type	**Te/MW**	**m^3/MW**	**% Efficiency**	**Duty Cycle**	**Notes**
High speed diesel mechanical	3.53	3.69	44 (m)	Heavy	Engine only
Medium speed diesel mechanical	4.56	28.56	46 (m)	Heavy	Engine only
Gas turbine mechanical	0.84	4.08	40 (m)	Heavy	Module only
HT-PEM, automotive	18.70	9.23	50 (e)	Light	Complete system [16]
Generic automotive SOFC	1.00	1.00	60 (e)	Light	Stack only [17]
SOFC (Domestic, CHP)	125.00	437.00	52 (e) 90 (CHP)	Medium	Complete system [18]
SOFC-GT hybrid (aircraft)	3.43	3.96	70.4 (e)	Heavy	Complete system [19]
SOFC-GT hybrid (Industrial, CHP)	145.24	619.96	55 (e) 73 (CHP)	Heavy	Complete system [20]
	Te/kVA	**m^3/kVA**			
High speed diesel generator	8.08	7.52	44 (e)	Heavy	Engine and alternator

4.3 FUEL CELLS

Hybrid SOFC-GT arrangements, where a gas turbine is driven from the hot exhaust of the fuel cell, are capable of high efficiencies, near 60% with simple cycle gas turbines and up to 70% if complex cycle machines are used. An alternative arrangement to achieve very high overall efficiency is to use waste heat for domestic or process heating, in a Combined Heat and Power (CHP) arrangement. Whilst this is common on commercial vessels (e.g. the use of engine cooling water for flash distillation), much like the complex cycle gas turbine, it is best suited to land applications where space and weight are not an issue and loads are predictable or slow changing. This is apparent in the very large values for weight and volume per unit power for the industrial and domestic SOFC hybrid options.

Fuel cell technology is being driven by other industries; land-based for reduced emissions from natural gas, automotive or aerospace for the more electric aircraft. The use of aerospace derivative fuel cells is of particular interest as unlike automotive systems, they are designed for a high duty cycle, with the demand changing comparatively little through flight [19]. This better matches the loading seen by marine machinery, particularly if battery systems can be used to provide peak power for short periods.

Where fuel cells were used in the studies described in this paper, battery systems providing 30 minutes of cruise power and propulsion loads were included, compared to 30 minutes of hotel load only in non-fuel cell options. This is an assumed duration, with the exact sizing of energy storage in fuel cell systems being a detailed subject. As was noted in the submarine case in Pawling, 2021 [21], lithium-ion batteries cannot be packed into battery tanks as is the case with lead-acid cells in submarines. Instead they require enclosures containing control systems, cooling, fire protection etc. which increases volume requirements significantly.

5. OPV STUDIES

5.1 BASELINE DESIGN

The starting point for the ongoing studies was an Ocean Patrol Vessel (OPV). The requirements for the baseline OPV are summarised in Table 5 below. Fuel tankage is determined by a two-week operational profile as shown in Figure 4 below, with the resulting cruise speed fuel endurance being an output, but generally 25-35 days depending on prime mover type.

Table 5: Summary of OPV design requirements

Performance		**Mission Systems**	
Cruise Speed	14 knots	**Weapons**	1 x 76mm gun 2 x 30mm guns
Maximum Speed	24.5 knots	**Aviation**	1 x 5te helicopter with hangar
Stores Endurance	28 days	**Daughter Craft**	2 x 7m RIBs
Accommodation	75	**Sensors**	1 x 2D surveillance radar 2 x nav radars 1 x electro optical 1 x fire control radar

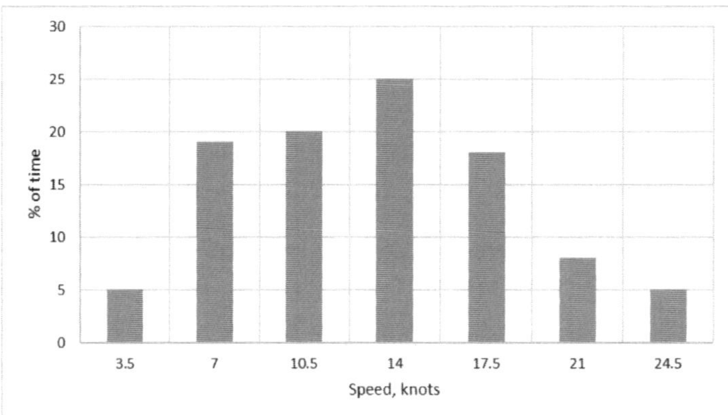

Figure 4: OPV design operating profile

The power and propulsion system for the baseline OPV design is shown schematically in Figure 5. A twin-shaft CODLAD arrangement was adopted, with diesel generators providing power for hotel loads and cruise speed propulsion, with diesel engines for boost to maximum speed. High-speed motors running at a nominal 1000rpm and connected to the gearboxes were used, as geared high-speed motors have been shown to lead to weight savings [23]. Variants retained this broad topology, with prime movers being replaced with fuel cells.

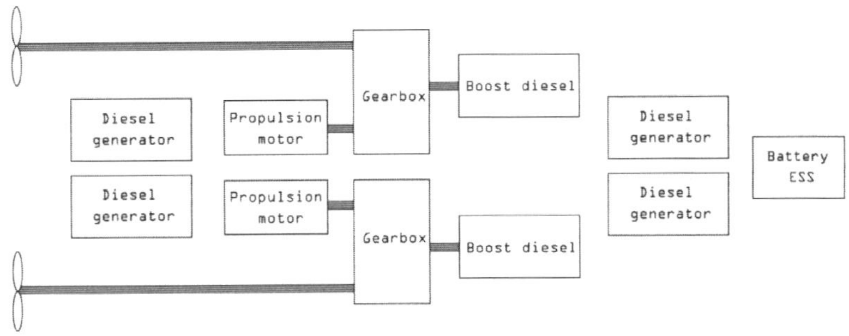

Figure 5: Baseline design machinery sketch

The profile distribution of the four main Functional Groups, FLOAT, MOVE, FIGHT and INFRASTRUCTURE are shown in Figure 6 below. FLOAT volume is dominated by the boat bays and hull voids, whilst MOVE is concentrated amidships with the motor and boost diesel machinery rooms adjacent, bookended by AMRs containing the diesel generators (INFRA). FIGHT shows the aviation spaces aft, command complex in the superstructure and 76mm gun forward.

float

	Stern H	G	F	E	D	C	B	Bow A
SS H	0	0	0	259	8	0	0	0
SS L	0	0	34	261	0	7	0	0
No 2	231	3	2	8	60	13	34	35
H L	0	46	83	58	68	92	105	136

move

	Stern H	G	F	E	D	C	B	Bow A
SS H	0	0	7	0	7	222	0	0
SS L	0	0	7	37	7	0	0	0
No 2	0	0	13	6	68	2	0	0
H L	131	33	144	376	281	221	0	0

fight

	Stern H	G	F	E	D	C	B	Bow A
SS H	0	0	248	0	0	138	0	0
SS L	0	0	268	0	155	124	0	0
No 2	0	0	0	0	0	0	69	0
H L	0	32	0	0	0	0	66	0

infra

	Stern H	G	F	E	D	C	B	Bow A
SS H	0	0	8	4	86	86	0	0
SS L	0	1	8	4	134	142	0	0
No 2	4	280	277	325	148	191	43	6
H L	0	39	123	3	42	82	3	0

Figure 6: Functional group volume distribution for the baseline design

Running ZEOLIT with a +/-5% balance limitation resulted in a spread of designs from 1700 to 2300 tonnes, with two peaks representing two common displacements as shown in Figure 7. The lowest value of circular M, 6, was the most common, along with low B/T (3) and high draught to depth ratio (0.5). higher values of B/T would be required to meet stability requirements, however, as shown by Figure 8. The overall size compares well with another helicopter equipped OPVs.

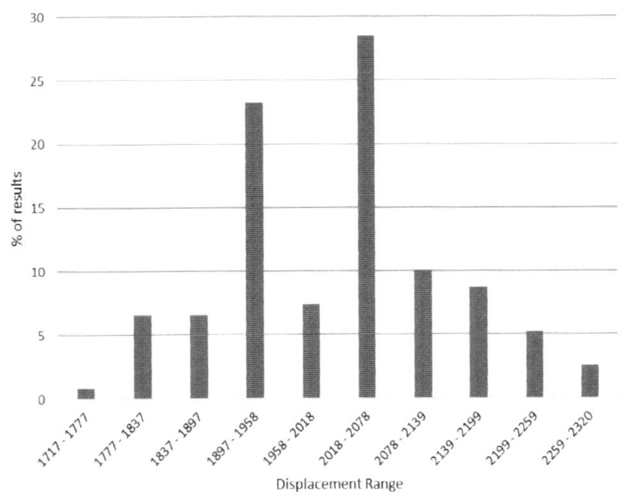

Figure 7: Displacement distribution for baseline design

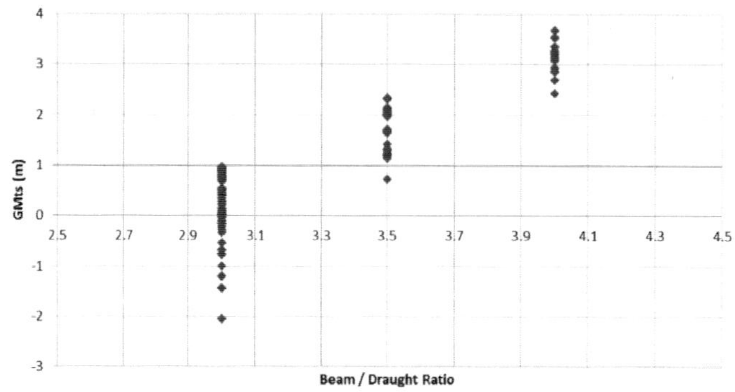

Figure 8: Variation in estimated GMts with B/T ratio, based on hullform characteristics and KG derived from overall layout

5.2 METHANOL FUEL

5.2 (a) Methanol Internal Combustion Engine (ICE)

The first variant to be investigated retained the use of methanol as a fuel in internal combustion engines. For this initial study, no physical changes to the engines were modelled. Figure 9 compares the baseline profile with the methanol-ICE variant, and Figure 10 shows the change in displacement distribution in the results.

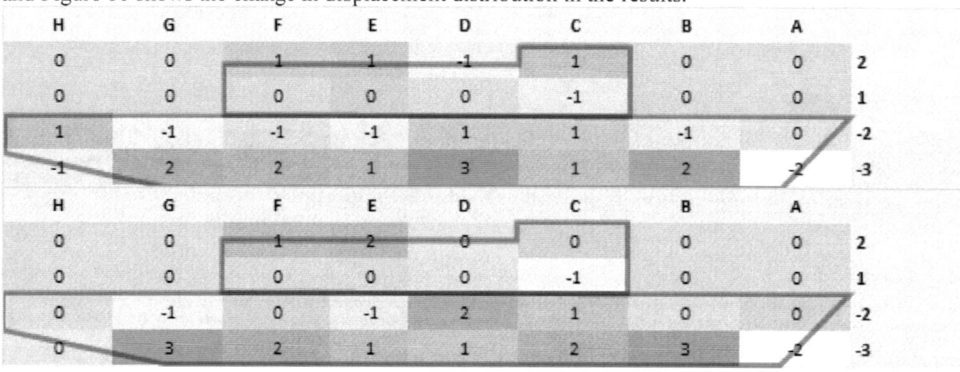

Figure 9: OPV baseline (top) and methanol ICE variant (bottom); higher numbers indicate more options have excess volume in that region of the ship

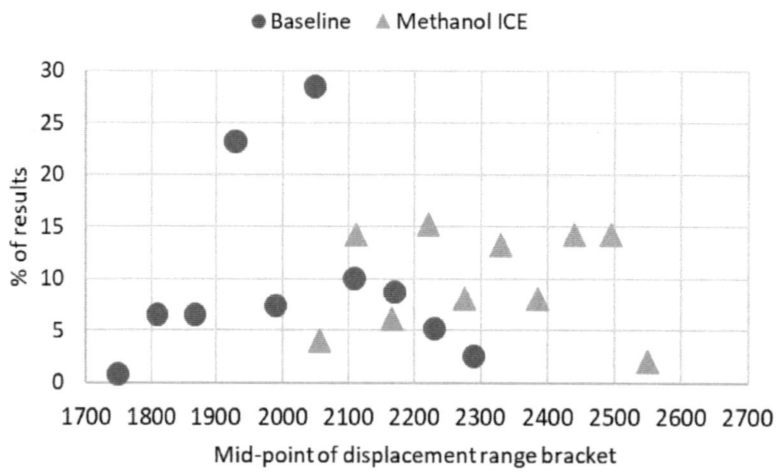

Figure 10: Comparison of displacement distributions for the baseline and methanol-ICE variant

It can be seen that the use of a less energy dense fuel led to an increase in ship size, with the most common displacement range increasing from the 2018 – 2078tonne bracket to 2193 – 2248tonnes. Although the total variation in displacements was similar (603 vs 548 tonnes), the methanol variant was not dominated by any particular size range.

Examining the arrangement aspect, Figure 9 reveals that, contrary to expectations (given the increased size of the fuel tanks), the hull was actually more spacious than the baseline. This is due to the size of the items such as machinery and accommodation etc. only changing slightly with the small increase in hull size. The primary change in the arrangement was to move the deep tanks from a position forward of the machinery to a large tank separating the forward main and auxiliary machinery spaces, as illustrated by the change in fuel tank distribution shown in Figure 11 below.

	Stern							Bow
	H	G	F	E	D	C	B	A
SS H	0	0	0	0	0	0	0	0
SS L	0	0	0	0	0	0	0	0
No 2	0	0	0	0	0	0	0	0
H L	0	0	0.1	0.1	0.3	0.5	0	0
1	*0*	*0*	*0.1*	*0.1*	*0.3*	*0.5*	*0*	*0*

	Stern							Bow
	H	G	F	E	D	C	B	A
SS H	0	0	0	0	0	0	0	0
SS L	0	0	0	0	0	0	0	0
No 2	0	0	0	0	0	0	0	0
H L	0	0	0.1	0.1	0.55	0.25	0	0

Figure 11: Fuel tank distributions for the baseline (left) and methanol-ICE variant (right)

5.2 (b) Methanol HT-PEM Fuel Cell

The next variant replaced the ICE diesel generators with high temperature PEM fuel cells, with physical parameters based on the Blue World Technologies devices [16], assuming that the physical size of the fuel cell will not be impacted by increasing the duty cycle. Boost engines remained diesels running on methanol. The resulting volume availability and displacement distribution are shown in Figures 12 and 13 below.

H	G	F	E	D	C	B	A	
0	0	1	1	0	1	0	0	2
0	0	0	0	0	-1	0	0	1
0	-1	0	0	2	1	0	0	-2
0	3	0	2	2	2	3	-2	-3

Figure 12: Volume availability distribution for the methanol HT-PEM variant

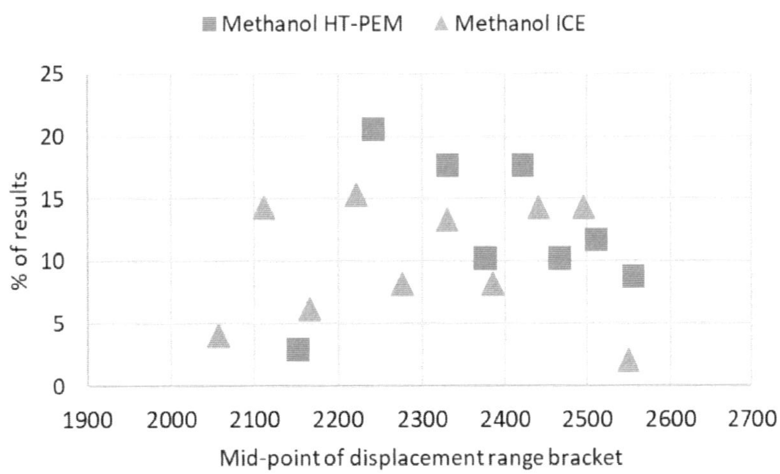

Figure 13: Comparison of displacement distributions of the methanol-ICE and methanol HT-PEM variants

Comparing Figures 12 and 9 it can be seen that this variant has more space in the hull. This is due to a combination of the improved fuel efficiency and increased weight of the fuel cells compared to the ICE engines. The modal displacement range actually increased, shifting from 2193 – 2248tonnes to 2219 – 2264. The larger hulls were no longer full of methanol, instead being needed to provide buoyancy to support heavier prime movers.

5.2 (c) Methanol SOFC-GT Hybrid

Retaining the hybrid fuel-cell / ICE machinery topology as the previous variant, the HT-PEM fuel cells were then exchanged for high temperature SOFCs with a GT waste heat recovery / reheat stage, based on studies for aviation APUs, with the efficiency set to 60% [18]. This had a lower modal displacement range, 2125 – 2189 tonnes, and less excess space in the lower hull.

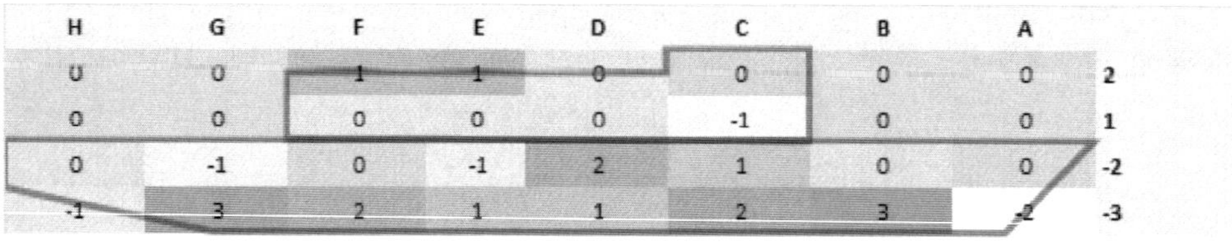

Figure 14: Volume availability for the methanol SOFC-GT variant

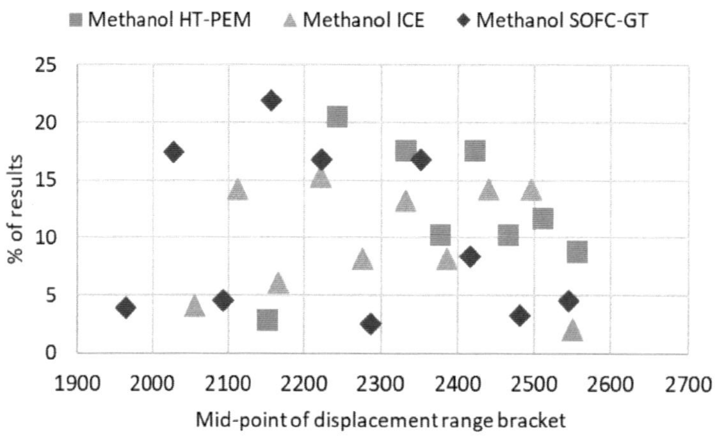

Figure 15: Comparison of all the methanol-fuelled variants

5.3 AMMONIA FUEL

The next variant retained the hybrid SOFC-GT and ICE machinery arrangement but switched the fuel to ammonia stored in a single tank block amidships. As can be seen in Figure 16, this led to the midships being cramped whilst the ends of the ship were relatively spacious due to an increase in displacement needed to accommodate the less dense fuel.

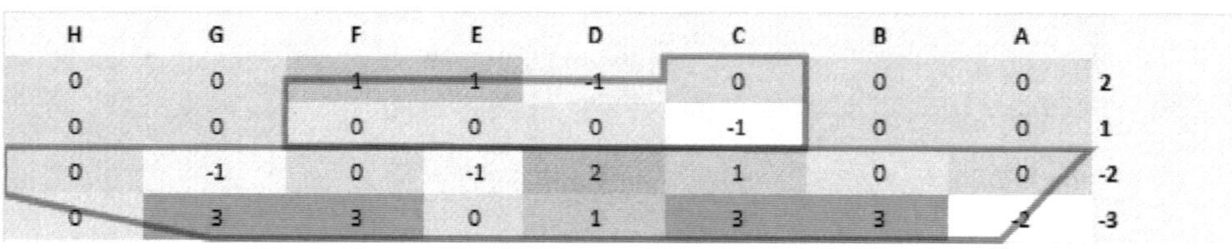

Figure 16: Volume availability plot of the ammonia fuelled variant, showing spacious lower hull

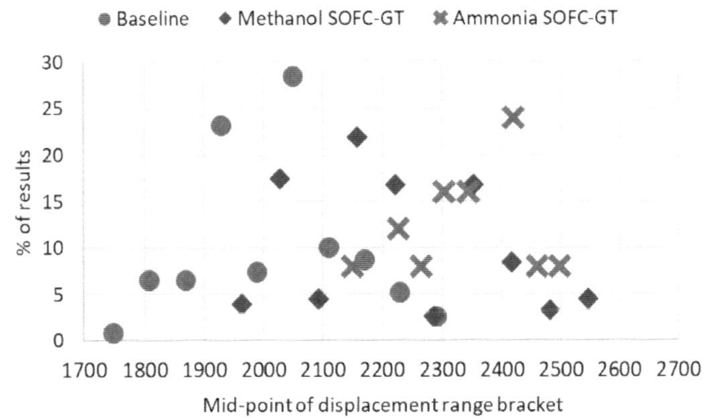

Figure 17: Comparison of F-76 fuelled baseline, methanol SOFC-GT and ammonia fuelled variants

With a modal displacement range of 2400 – 2439 tonnes, this was the largest variant yet, with the fuel choice driving the design. It should be noted that, although it was assumed that crew complement would not change for these variants, other spaces and weights such as distributed systems, hotel loads etc, did scale with the increased size of the ship.

5.4 LOHC

The final variant developed to date is an all-electric option fuelled by the benzyltoluene /perhydro benzyltoluene combination. Early iterations showed that using this fuel for the entire speed range would cause weight and space balance to be unachievable within the +/-5% limits (meaning that large additional void volumes are required) and so LOHC supplied hydrogen is used for cruise speed propulsion and hotel load generation, with methanol being used for the sprint speed. As Figure 18 shows, the resulting ship was very spacious, as the displacement was driven by the heavy machinery and large amounts of fuel, rather than the relatively modest accommodation and payload. It should be noted that the effective specific fuel consumption of an LOHC (remembering that the dehydrogenated product remains onboard) is in excess of 1kg/kWhr.

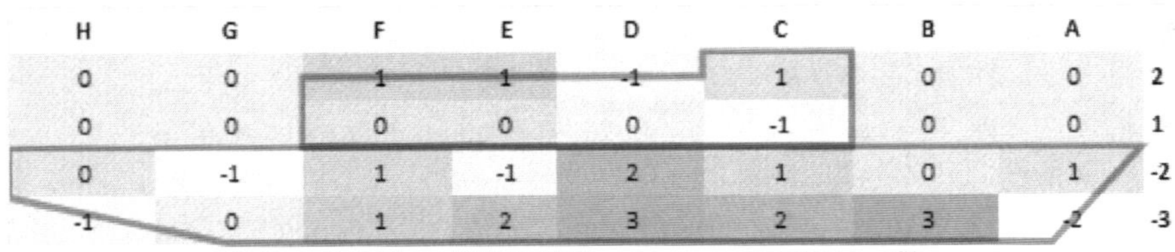

Figure 18: LOHC volume availability profile

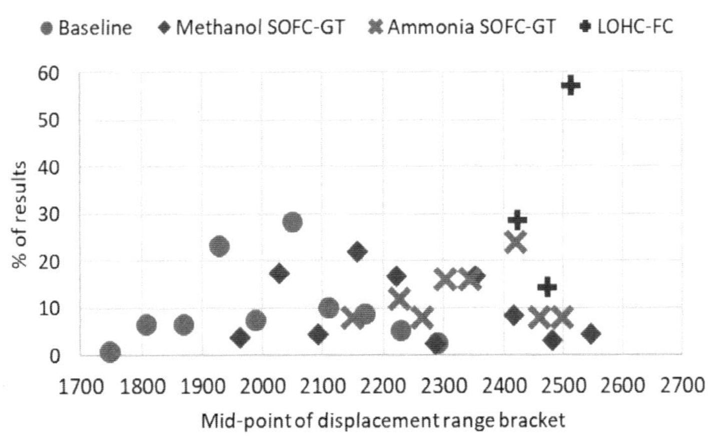

Figure 19: Comparison of baseline, best methanol, ammonia and LOHC variants

With a modal displacement range of 2508 – 2518 tonnes, the LOHC variant was the largest, but, due to the dual-fuel approach adopted, not outside the range for OPVs.

5.5 OVERALL COMPARISONS

Table 6 below summaries some key comparison metrics. The most obvious conclusion is that the use of alternate fuels, with their lower energy density, generally increases the size of the ship. It can also be seen that fuel cells, whilst reducing the amount of fuel required (represented by the fuel fraction; i.e. how much of the deep displacement is fuel) do not reduce the size of the ship as their increased size, and in particular weight, counteract this.

Consideration of the fuel fraction parameter leads to another aspect; with greatly increased fuel tankage comes increased changes in draught and trim as the fuel is used, with an increased rise in the KG. For the LOHC fuel, the opposite is true – "light" displacement will be close to "deep" displacement, which in itself may increase fuel burn as ships inherently have to spend most of their life at lighter displacements, as illustrated in [24].

Table 6: Selected comparison metrics

Variant	Baseline	Methanol ICE	Methanol HT-PEM	Methanol SOFC-GT	Ammonia SOFC-GT	LOHC
Modal Displacement Range (tonnes)	2018 - 2078	2193 - 2248	2219 - 2264	2125 - 2189	2400 - 2439	2508 - 2518
Average Fuel Fraction	0.110	0.204	0.188	0.180	0.173	0.267
Max. delta-T, (m)	0.384	0.707	0.664	0.672	0.609	n/a

The inclusion of a simplified arrangement in ZEOLIT illustrated that future fuels may in fact lead to increased void volumes in OPVs, contrary to the authors original expectations (given the larger volume required by the fuel). This is illustrated in Figure 20 below, which compares the weight and space balances for the baseline, methanol SOFC-GT and LOHC variants. Values >1 mean excess volume or buoyancy to the hull reference waterline, whilst values <1 mean insufficient volume or buoyancy. Note that a +/-5% allowance was used in all studies. Figure 20 shows that the volume becomes increasingly excessive as the fuel becomes less energy dense, as the volume balance points (blue diamonds) are increasingly concentrated in the upper part of the plot.

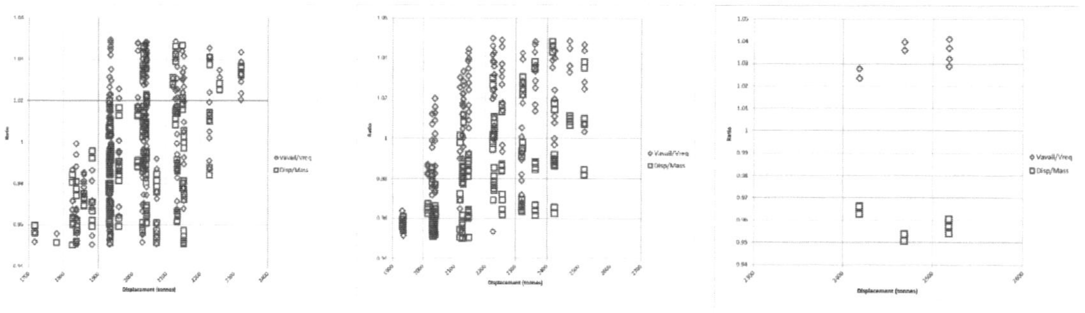

Figure 20: Changes in the distribution of available/required volume (blue diamonds) and buoyancy/weight (red squares) with reducing fuel energy density

6. CONCLUSIONS AND FUTURE WORK

6.1 THE INITIAL STUDY

This paper as described the first stage in an ongoing study to examine the impacts of possible future fuels on the design of a range of naval vessels. The use of the UCL ZEOLIT tool reflects a desire to keep the modelling at a high-level and capture overall trends, to develop general guidance for future ship designs. The initial stage has examined an OPV, as this is a relatively simple type of vessel, which has already seen the adoption of dual fuelling with LNG in some navies. Three future fuels have been investigated: methanol, ammonia and a Liquid Organic Hydrogen Carrier.

The findings so far indicate that, for an OPV, these fuels lead to an increase in displacement, but not one that would render the ship impractically large. However, dual fuelling was required for the least energy dense fuel (LOHC) to avoid the high-speed operating mode dominating the design. Another finding relevant to future designs is that less energy dense fuel types, in combination with heavier machinery such as fuel cells, will change the balance of volume within the hull, with more void space (or more spacious compartments) resulting.

6.2 FUTURE WORK

6.2(a) Marine Engineering Definition

This initial study was conducted by the author to determine if the results indicated the subject being worth further investigation. The next stage will involve collaboration with marine engineering colleagues in UCL to improve the definition of the machinery plant. In particular, the conversion of fuel cell data derived from other industries into maritime configurations, and a better representation of the balance of useful and waste energy for the high temperature fuel cells.

6.2(b) Additional Ship Types

The current hull databased used by ZEOLIT is limited to a displacement between 400 and 4300 tonnes – i.e. OPVs, corvettes and light frigates. Work is underway to expand this to 10,000 tonnes to allow assessment of frigates and destroyers. Above this size, however, the division of the ship into 32 regions may be too crude – specifically in the number of "deck" slices.

6.2(c) Costing

One area of future work will be the addition of cost estimates. UCL has a weight and equipment-based costing model with bespoke coefficients, that generally performs well for the overall cost of a warship assuming no major programmatic problems. Figure 21 compares the cost per tonne for various real-world vessels and similar UCL student designs.

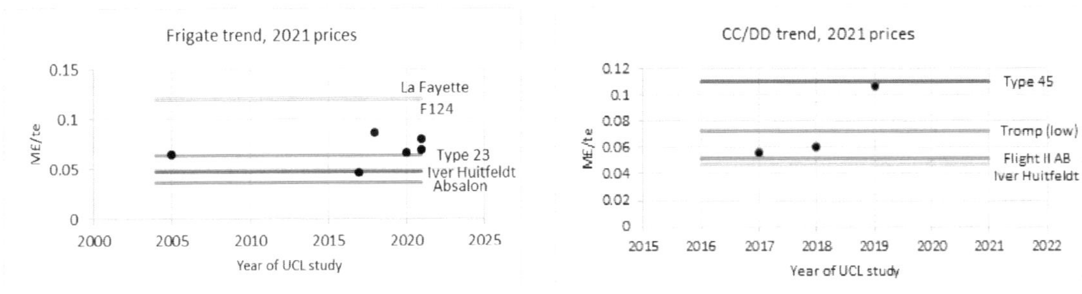

Figure 21: Comparison of UCL MSc costs per tonne (points) with real-world public-domain cost values (lines)

However, obtaining cost values for technologies such as fuel cells is difficult, as data rapidly becomes out of date, as shown in Figure 22 which illustrates the change in the cost estimates for automotive PEMs, based on detailed parametric modelling. This reduction in cost has been driven by similar factors to the drop in cost of lithium-ion batteries: changing raw material costs; improvements in manufacturing technologies; increased production numbers.

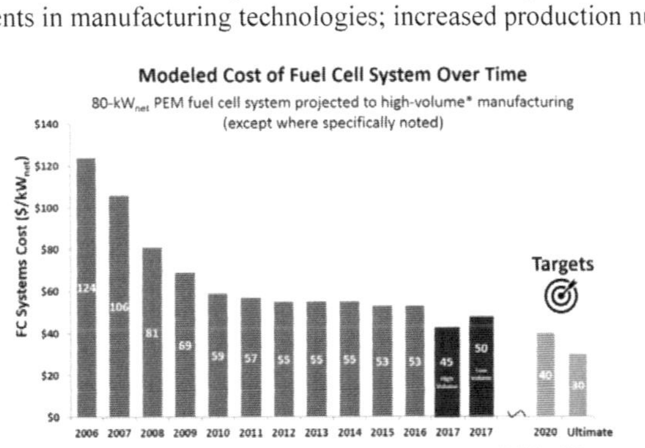

Figure 22: US government bottom-up modelling of fuel cell production costs showing significant reductions [25]

Whilst the values shown in Figure 22 make fuel cells seem very attractive, they are for light-duty automotive PEMs using hydrogen fuel. Adding fuel processing, increasing temperature (so allowing auto-reformation) and increasing the duty cycle all increase unit cost. Increasing the duty cycle from light (a car) to medium (a truck) will increase fuel cell system cost by a factor of 4 [15] and increasing to heavy duty (i.e. continuous running) will increase this cost further. However, at present this may be a factor of fuel cell research having been focussed on the automotive side, so the developments have not yet propagated to other use cases.

7. REFERENCES

1. Pawling, R. J., "The Zoom Years: UCL MSc Student Warship Designs Through the Pandemic", RINA Warship 2022, 8th – 9th June 2022, Bristol, UK

2. Pawling, R.J., Farrier, L., Bradbeer, N., "The Advanced Technology Corvette (Railgun): A Concept Exploration of Future Warships", International Conference Warship 2018: Procurement of Future Surface Vessels, RINA, London, June 2018.

3. Pawling, R.J., Farrier, L., Bucknall, R.W.G., "The Advanced Technology Corvette - Railgun (ATK-R) Design Study - Future Weapons and Small Ship Power Systems", Proc. International Naval Engineering Conference (INEC) 2018, Institute of Marine Engineering, Science and Technology (IMarEST), Glasgow, UK

4. McKesson, C., "The Utility of Very Simple Models for Very Complex Systems", SCS M&S Magazine – 2011 / n1 (January)

5. DNV-GL, Comparison of Alternative Marine Fuels, Report No.: 2019-0567, Rev. 3, July 2019

6. Krants, R., Søgaard, K. & Smith, T., "The scale of investment needed to decarbonize international shipping", Global Maritime Forum, https://www.globalmaritimeforum.org/news/the-scale-of-investment-needed-to-decarbonize-international-shipping

7. DNV, "Alternative Fuels for Naval Vessels", white paper, 2022

8. Astley, W. E., Grasman, A. & Stroeve, D. B., "Exploring the Impact of Methanol as an Alternative, Cleaner Fuel for the Auxiliary and Support Vessels Within the RNLN", Proc. INEC 2020, IMarEST, October 2020

9. Tol, R. & Linden, Y., "Effect of Gas-to-Liquid (GTL) Fuels on Marine Diesel Engines Compared to F-76", Proc. INEC 2018, IMarEST, October 2018

10. Alberico, E. & Nielsen, M., "Towards a methanol economy based on homogeneous catalysis: methanol to H2 and CO2 to methanol", Chem. Commun., 2015, 51, 6714-6725

11. Rude, T., Durr, S., Preuster, P., Wolf, M. & Wasserscheid, P., "Benzyltoluene/perhydro benzyltoluene – pushing the performance limits of pure hydrocarbon liquid organic hydrogen carrier (LOHC) systems", Sustainable Energy Fuels 2022, 6, 1541

12. Brack, D., Birdsey, R. & Walker, W., "Greenhouse gas emissions from burning US-sourced woody biomass in the EU and UK", Chatham House

13. Banks et al., "Understanding ship operating profiles with an aim to improve energy efficient ship operations", Low Carbon Shipping Conference, London 2013

14. Pawling, R.J., Suárez de la Fuente, S., Andrews, D. J., "The Potential Use of Energy Saving Technologies in Future Patrol Combatants", International Conference Warship 2016: Advanced Technologies in Naval Design, Construction and Operation, RINA, Bath, June 2016

15. Advanced Propulsion Centre UK, Fuel Cell Roadmap 2020, ISBN 978 1 78413 493 8

16. Jensen, M. F., "Blue World Technologies", Methanol Vehicles Exhibition, Chongqing, October 2019

17. Bossel, U, "Solid Oxide Fuel Cells for Transportation", Journal of KONES Internal Combustion Engines 2005, vol. 12, 3-4

18. Kyocera: "KYOCERA Develops Industry's First 3-Kilowatt Solid-Oxide Fuel Cell for Institutional Cogeneration", July 6, 2017, https://global.kyocera.com/news/2017/0702_bnfo.html

19. Whyatt G.A. & Chick L.A.: "Electrical Generation for More-Electric Aircraft using Solid Oxide Fuel Cells", PNNL-21382, US DoE, April 2012

20. Mitsubishi Power Sustainable Energy Solutions, Solid Oxide Fuel Cell System, Mitsubishi Power Europe GmbH, 07.07.21

21. Pawling, R., Dos Santos, J., Oguz, Y., Weatherall, D. & Lango, B., "RENEWABLE: An Innovative UCL Concept Submarine Design for an All-electric Boat", RINA Warship 2020, 17th – 18th June 2020, Bristol, UK

22. Chowaniec, Z., Forbes, J. & Shaw, J., "High speed motor propulsion for warships", Proc. INEC 2012, IMarEST

23. Lyster, C. & Pawling, R. "A proposed framework for developing an Energy Efficiency Design Index (EEDI) for Warships", MARI-TECH 2019

24. https://www.energy.gov/eere/fuelcells/fact-month-april-2018-fuel-cell-cost-decreased-60-2006

8. AUTHORS BIOGRAPHY

Dr. Rachel Pawling is a Lecturer in Ship Design at UCL, teaching the subject of ship design to undergraduate and postgraduate students. She obtained her PhD in computer aided ship design in 2007 and has subsequently continued her research in projects funded by the EC, UK and US governments.

AN EVALUATION OF A COATING SYSTEM THAT DELIVERS TOP SPEEDS, SUPERIOR MANOEUVRABILITY AND LONG-RANGE CAPABILITIES FOR THE SUSTAINABLE FUTURE OF WARSHIPS

M N van Ruiten, Subsea Industries, The Netherlands

SUMMARY

Coating systems are used in the shipping industry to protect the vessel against corrosion. The other main goal of coating systems is providing optimal hull performance; for warships consisting out of top speed, long-range capabilities, and superior manoeuvrability. This can be achieved by preventing biofouling and having a smooth hull without paint degradation. However, with current systems, biofouling still has the change to attach to the ship's hull despite the fact that these coatings have a significant environmental impact. These coatings will last for 3-5 years before it must be replaced as paint degradation develops over time. This results in a life cycle of re-applicating the coating every drydock session and sailing with excessive drag reduction, which involves significantly on the performance, fuel consumption, material resources, associated transport, and overall costs. An alternative ideal for warships, named Surface Treated Composites (STC), is investigated in this paper. STCs are designed for regular in-water cleaning throughout the lifetime of the coating without the need for reapplication. Regular in-water cleaning results in a significant drag reduction as the hull can be smooth whenever is needed. STCs therefore offer a viable antifouling strategy that is entirely non-toxic. The application of STCs leads to top speeds, long-range capabilities, superior manoeuvrability, and lower greenhouse gas emissions. This provides in financial benefits for the shipowner or shipping company because fuel is saved greatly and there are no expenses associated with repainting every couple of years. As a result, the STC coating is not only ecologically friendly but also cost-effective.

NOMENCLATURE

AF Antifouling:
Antifouling paints contain toxic substances designed to kill the fouling before it attaches to the ship's hull or after it has attached. They work based on leaching poisons continually as a way of dealing with biofouling.

FR Foul Release:
Fouling release coatings are designed to prevent fouling by surface effects and are not dependent on the traditional polishing action. However, these coating systems are dependent on certain minimum speeds for the fouling to be released, and the coatings are more subject to mechanical damage and roughening than traditional AF coatings. These coatings also contain toxic substances that are harmful to aquatic organisms.

IMO International Maritime Organization:
A specialized agency of the United Nations that is responsible for measures to improve the safety and security of international shipping and to prevent marine pollution from ships.

NIS Non-indigenous Species:
An introduced, alien, exotic, non-indigenous, or non-native species, is a species living outside its native distributional range, which has arrived there by human activity, either deliberate or accidental. Some introduced species are damaging to the ecosystem they are introduced into and then reach invasive status, negatively affecting the local ecosystem.

STC Surface Treated Composite
STC is made up of glass platelets mixed in a resinous base. The result is a very hard, long-lasting coating. An STC is entirely non-toxic. The hard smooth surface is ideal for in-water cleaning, particularly at the slime or light weed stage, but even if heavily fouled it can be cleaned and restored to its original condition without damage or deterioration. The concept of surface treated composites is patented.

VOC Volatile Organic Compounds
Any carbon compound that evaporates under standard test conditions. Essentially, all paint solvents except water are VOCs. Federal and state governments are beginning to limit the number of volatile organics found in paint because of concerns about possible environmental and health effects. VOCs are an element of smog and poor air quality.

dft Dry film thickness (μm)

Dry film thickness is the thickness of a coating as measured above the substrate. This can consist of a single layer or multiple layers. DFT is measured for cured coatings (after the coating dries).

1. INTRODUCTION

Coating systems have two main purposes: to protect the vessel against corrosion, and to ensure optimal performance by keeping the surface of the hull smooth and frictionless. Top speed, long-range capabilities, and superior manoeuvrability are all components of optimal performance for warships.

To ensure optimal performance, the underwater hull must be protected against the attachment of marine organisms, and it should maintain a smooth hull over time without paint degradation. The accumulation of algae and barnacles on the ship's hull has a negative impact on the hydrodynamics of the warship and results in a significant decrease in performance.

Recent research of the IMO shows higher-than-expected fuel cost from fouling [1]. And according to Svitzer [2], shipping companies lack the focus on holistic hull performance. There is a poor understanding of the immediate impact that good hull performance can have on fuel consumption and emissions, particularly compared to its relative simplicity and cost-effectiveness.

Antifouling (AF) and foul release (FR) coatings have traditionally been used to prevent biofouling. The most common coating in the shipping industry is an AF coating, which gradually leach biocides through a chemical reaction and have a lifetime of up to 5 years [3]. The continued leaching of biocides into the aquatic environment raises serious concerns. The input of copper into water from AF paints has been conservatively estimated to be around 5000 tonnes per year, with the main concern being the localised scale of input in enclosed water bodies such as marinas, ports, and harbours [4].

FR coatings are a "non-stick" type of coating which works on the principle that it is difficult for fouling to stick to it in the first place and easy for it to fall off, wash off or "release" when the vessel is under way, especially at speed [3]. The coatings currently available in this category are silicone-based and have, just as the AF coatings, a negative impact on the environment.

The current approach, that consists primarily out of these two coating systems, is analysed, and discussed in the next section. Thereafter, the STC approach will be analysed. This approach is a 100% non-toxic, hard, inert glass flake coating referred to as a Surface Treated Composite (STC) that deals with biofouling by cleaning the underwater part of the vessel.

2. CURRENT APPROACH

The current approach, consisting out of an AF and a FR coating, is analysed by the requirements for a theoretically ideal coating system. The theoretically ideal coating should guarantee that:

- It maintains a smooth hull
- No attachment of marine organisms occurs
- It protects the ship over its entire lifetime
- It is completely safe to the environment

In the following subsections, the current approach is analysed and discussed based on the four requirements indicated above.

2.1 PAINT DEGRADATION

Warships should have a smooth hull during operations to ensure top speed, long-range capabilities, and superior manoeuvrability.

Due to the nature of the AF and FR coatings used, the complex build-up of different layers of different substances on a substrate which naturally interacts electrolytically with the metals in the biocide coatings leads to significant degradation of the coating. Besides, spot blasting, patching, and coating repair accelerates this degradation process (Hydrex Group, 2012).

The ship's hull becomes rougher and rougher over its life. Long before the hull is reblasted and a full new coating system applied, the fuel penalty caused by the added roughness, even without considering hull fouling, has been shown to reach 25 - 40%.

This means that more fuel is being burned than necessary to maintain speed. This is accompanied by a proportionate unnecessary emission of CO_2, NOx, SOx and particulate matter.

In Figure 1, the increase in average hull roughness and powering with ship age is presented.

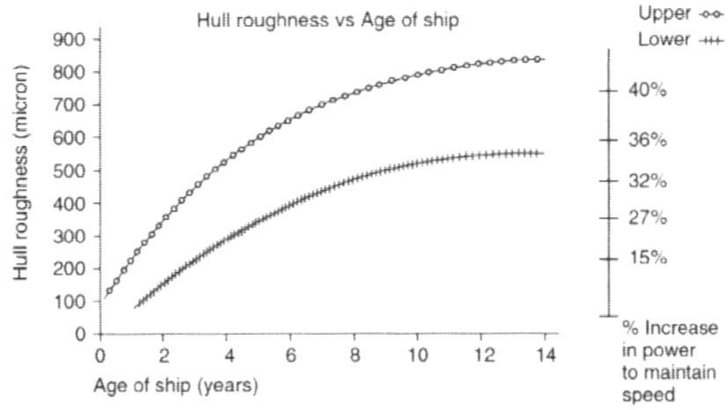

Figure 1 Increased average hull roughness with ship age and effect on powering [5]

2.2 BIOFOULING

Currently available AF and FR coating systems are not effective in preventing a build-up of slime. This is despite the chemicals used. In fact, all available coatings tend to accumulate biofouling in the form of slime or biofilm quite rapidly, depending on the ship's sailing pattern and the waters in which it sails (Hydrex Group, 2011).

2.2 (a) Low performance and manoeuvrability

Compared to no fouling control, an antifouling coating which is effective in deterring or killing macrofouling organisms, will prevent the huge fuel penalty which accompanies heavy fouling. This will result in lower fuel consumption than that of a vessel with heavy fouling. However, these coatings tend not to be effective in preventing a slime layer from building up. Since slime can carry a 20% fuel penalty, these vessels do not perform at their optimal performance, they have a reduced range and the vessels are poorly manoeuvrable [6].

In Figure 2, the relation between the additional power required and the hull fouling rate at two different speeds is presented.

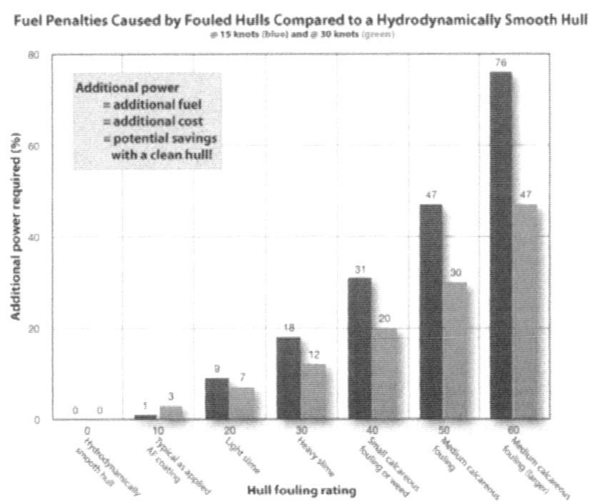

Figure 2 Fuel penalties caused by fouled hulls (Schultz et al., 2010)

3.2 (b) GHG-emissions

More severe biofouling conditions can lead to higher emissions, showing the importance of good biofouling management. The impact biofouling has on the increase in GHG emissions is presented in the graph provided by the IMO, visible in Figure 3.

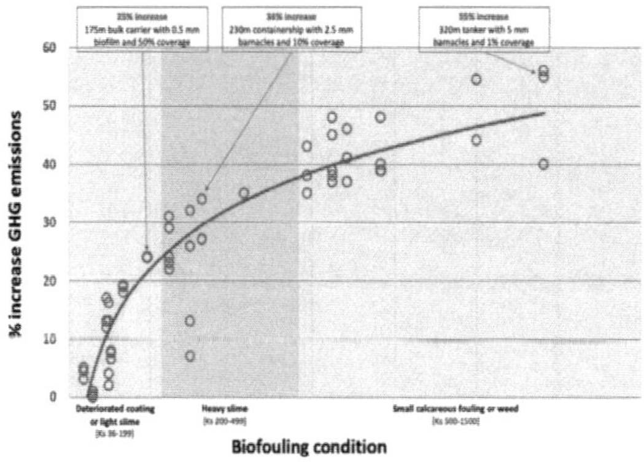

Figure 3 Excessive GHG emissions is the price to pay [1]

The early findings of IMO's study on the impact of biofouling reveals that a minor layer of slime as thin as 0.5 mm covering up to half of a hull surface can trigger an increase of GHG emissions in the range of 20 to 25%, depending on ship characteristics, speed, and other conditions [1].

With a light layer of barnacles or tubeworms, an average container ship can see an increase in GHG emissions of up to 55 percent, dependent on the ship's characteristics and speed [1].

As the biofouling conditions are not manageable with the AF and FR coatings today, excessive GHG emissions is the price to pay.

2.3 LIFESPAN

The AF and FR coating systems are classified as soft coatings. Typically, these will last 3 - 5 years before it must be replaced [3]. These times vary with the type of ship, the coating system used, the surface preparation prior to application, and the waters where the ship operates: for example, if the ship sails in Arctic waters, or if it is a pilot vessel operating in a harbour. These times can be regarded as general average.

When the AF properties are depleted, the ship must be drydocked, and the hull coating system repaired where damaged, which requires the full application of all coats to those areas, including replacing the anticorrosion layers. Then the entire hull needs to be repainted with one, two or more coats of biocidal antifouling paint.

Typical FR coating systems tend to be very easily damaged. Abrasion resistance is quite poor. Any collision or mechanical abuse of any kind is likely to damage the coating. The life cycle and reapplication pattern of a FR coating are like that of an AF coating. Repairs to FR coatings are more difficult than with AF coatings since they are non-stick coatings and there is invariably an area of overlap when a repair to a smaller area of the hull is required, and adhesion is a problem. The FR coatings are liable to the same under creep as the antifouling coatings, especially when the surface preparation prior to initial application was skimped.

Both coating systems fail to meet the primary reasons for applying a coating in the first place. These coatings are easily damaged and deteriorate over time. As a result, vessels are not adequately protected against corrosion.

2.4 ENVIRONMENTAL SAFETY

Next to the additional fuel consumption and excessive GHG emissions, there are several ways that the chosen hull coating can increase or decrease the ship's impact on the environment. These are, in essence,

- Chemical pollution of the marine environment during application, use, removal, or cleaning of the coating.
- Microplastic pollution
- Atmospheric pollution during the application of the coating by volatile organic compounds (VOCs) in the solvents.
- The spread of invasive non-indigenous species (NIS)

2.4 (a) Chemical pollution

While the ingredients and proportions and the method of release of biocide vary from one AF coating system to the next, they all have one thing in common: they work by releasing poisonous chemicals into the water. All antifouling coatings in use today emit poisonous chemicals into the water and these chemicals are in varying degree harmful to marine life, contaminate the environment and enter the food chain. This environmental pollution occurs constantly as toxic compounds are leaching into the water whenever the ship is at anchor or moored or sailing.

In FR coatings, the catalyst used is highly toxic as organotin and other biocides are included. It is therefore inaccurate to describe FR coatings in general as non-toxic or environmentally safe.

Both systems threaten the rivers, seas, and oceans including all their inhabitants. This chemical pollution enters the food chain and therefore pose a threat to public health.

2.4 (b) Microplastic pollution

Polymers are used for the resin of different layers for marine coatings; primer (epoxy), anti-corrosive layer and antifouling layer. As the paint particles are released into the water by the disturbance, erosion, and weathering of the coatings, microplastics (particles between 1 and 5 mm, depending on the definition) enters the oceans and rivers.

A new study published by the Swiss-based Environmental Action has found that 66% of the total paint used in the marine sector will eventually end up in the environment of which 90% will leak to ocean and waterways [7]. In total 1.9 million tons of paint found in oceans and waterways each year and outweighs other sources of microplastics, including textile fibers and tire dust.

The problem is that AF coatings are built to release into the water continuously, and that both AF and FR coatings are low abrasion resistant and do not attach well to the surface (eq. steel) and as a result, the polymers enter the environment in the form of microplastics. In terms of share of paint-related plastic pollution this represent 12% of the global amount.

2.4 (c) Volatile organic compounds

Volatile organic compounds (VOC) are organic chemicals (basically compounds containing carbon) with a high vapor pressure or low boiling point which makes it easy for them to evaporate from liquid form, for example, the solvents in paint which evaporate when the paint is applied and while it cures. Some VOCs are harmful to air quality, contributing to smog, and have compounding long- term ill effects on health. VOC levels of a typical AF coating are relatively more than those of typical FR coating systems (Hydrex Group, 2011). However, as both coatings need to be replied every 3-5 year, the VOC emissions are significantly high for both systems over the lifetime of a vessel.

2.4 (d) Non-indigenous marine species (NIS)

On the subject of non-indigenous species, AF coatings get a mixed review. Obviously if they kill or deter marine life it won't attach to the hull and therefore will not be translocated. However, there are species which are immune or develop immunity to the biocides in the AF coatings and when translocated these species have a competitive advantage over the aquatic species in their new home which are not so immune, and so can invade with greater chance of success (Hydrex Group, 2011).

For FR coatings, a high activity and higher speed vessels, it is unlikely that serious fouling will accumulate on the hull and therefore the NIS-issue will not materialize under these conditions. The potential problem arises when the vessel remains in port or at anchor for an extended period and fouling is permitted to build up. Such fouling may or may not wash off when the ship does sail. It depends if the fouling species have dug through the FR coating to the underlying anticorrosive scheme or to bare steel. If they have, then the likelihood of their washing off is low. Thus, non-indigenous species might be translocated [8].

2.4 (e) Overview conclusions

The current coatings do not allow warships to sail with the most optimal performance as paint degradation occurs and marine organisms still attach to the hull despite the toxic substances in the coating. This is even worse for ships who lay idle for a long time. The accumulation of biofouling and the development of paint degradation reduces the performance and protection of the ship. This results in additional maintenance, operational and environmental costs because the AF and FR coatings need to be re-applied frequently, fuel consumption increases, and it has a long-term negative impact on our environment. Therefore, a better alternative should be analysed.

3. SURFACE TREATED COMPOSITES

Having examined existing practices and concerns regarding underwater ship hull protection and fouling control and noted the shortcomings and issues, and compared the various coatings available, the next logical step is to describe a viable alternative technology and its practices in detail.

3.1 DESCRIPTION

An entirely different approach was introduced commercially in 2002 and has been in successful use since. Not just a coating, a Surface Treated Composite (STC) is a system of hull protection and fouling control comprising two main elements:

- A hard, inert, long-lasting, tough, ultra-smooth, glass-reinforced coating designed to last the life of the hull without need for replacement and with only minor touch-ups required during routine dry-docking.
- Routine underwater cleaning whereby the smoothness of the coating is improved over time and the hull is kept clean of any fouling beyond the light slime that will attach between cleanings.

3.2 STC COATING

One STC that is commercially available is formulated as a vinyl ester with a high concentration of embedded glass flakes, a schematic view of which is shown in Figure 4. This coating is typically applied in two coats of 500μm dft. each. The high film thickness and the presence of glass flakes which act as an impermeable barrier explain why this coating has excellent anticorrosive properties. It is extremely durable, making it excellently suited to regular in-water surface treatment and cleaning.

Tests have shown that a very large number of repeated underwater cleanings (500) on the same surface improve its surface smoothness and have no deteriorating effects [4]. The horizontal aligned glass flakes make it impossible for marine organisms, chemicals, and water to penetrate through the coating.

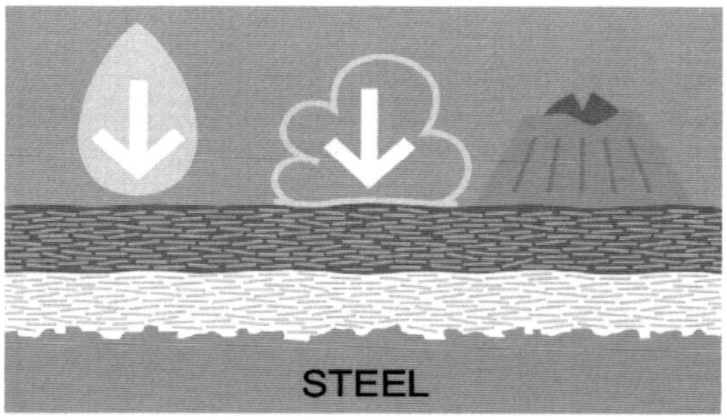

Figure 4 Built-up STC

3.3 DIFFERENT APPROACH

As the FR and the FR coating has been analysed based on the four requirements indicated in Section 2, the STC coating should be analysed the same way.

3.3 (a) Smooth hull

After a freshly applied coating, STC is hydrodynamically smoother than an antifouling coating. However, the FR coating shows to be slightly smoother [3].

After a couple of months, the AF and FR coatings degrade significantly. Whereas the STC coating shows to have a tough and flexible character, which causes that the coating does not degrade. This results in an ultra-smooth surface that can be maintained over the entire life of the vessel. Next to that, this coating is very corrosion resistant as it creates a barrier over the steel. Therefore, the coating lasts the lifetime of vessel with only minor touch-ups and no repaint required.

3.3 (b) Biofouling

A glass flake vinyl ester STC combined with routine in-water cleaning so that the ship never sails with more than light slime can deliver savings of 10-20% of fuel costs compared to AF or FR coated ships which will both build up a slime layer. In the case of ships with long layup times, the hard coating can be cleaned before the vessel sails, so that it sails with a clean hull, thus delivering the savings which, as stated, can be as much as 20% or, in the case of heavy fouling, a lot more.

The glass flake vinyl ester STC is very workable on ships that have lay-up periods providing the hull is cleaned before the vessel sails. It is best if the fouling is not allowed to develop beyond slime and weed, but even if it does, it can all be removed without any damage to the STC coating but rather the cleaning results in a smoother surface than before. This system permits very large savings in fuel even on ships that may be laid up for as much as a year before sailing again.

Figure 5 illustrates the difference in additional fuel consumption of a ship with layup times for a conventional coating and an STC coating. As can be seen in the figure, the STC coating may be cleaned to a perfectly smooth hull after each layup cycle. As a result, the vessel can sail with optimal performance during its operations. Biofouling and paint degradation occur with traditional coatings and are unmanageable in between dockings, leading to increased fuel consumption.

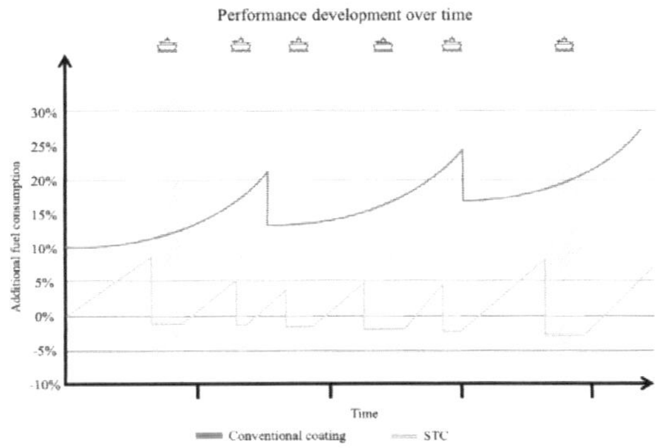

Figure 5 Development of performance over time of a conventional coating vs. STC coating

3.3 (c) Lifelong protection

Glass flake vinyl ester STC hard coating last the lifetime of the vessel with only minor touch-ups needed (typically less than 1% of the underwater hull surface) when the ship is in drydock routinely as required by class. This expectancy depends on initial standard surface preparation and application, but this is not difficult or complicated. The STC coating is extremely tough and durable while also being flexible. If any mechanical damage occurs, it is easily repaired since the 1000 microns dry film thickness (dft.) coating is entirely homogenous. Two coats of the paint are all that is required on bare steel (although three or even four coats can be used for special applications), and the curing time in between coats is 2 - 3 hours. Thus, the touch-ups can all be done in a day while the ship is in drydock for other reasons. This is the only known coating which requires no drydocking specifically for painting for the entire service life of the ship after initial application.

3.3 (d) Environmentally safe

Tests done on glass flake vinyl ester STC showed it to be entirely non-toxic to the marine environment when in use, or when cleaned in the water.

Glass flake vinyl ester resin, offer the best fuel savings and therefore lowest fuel consumption and GHG emission, because it can be cleaned as often as required in water with no adverse effects to the coating. It even shows a better hydraulic smoothness with routine cleanings. The hull can always be kept at a level of fouling no more severe than light slime and ships which have been stationary for some time can be fully cleaned in the water before they sail. Sailing with such a hull when it is kept clean can reduce fuel consumption by 20%, thus reducing GHG and other emissions by a similar factor.

The best way to ensure that non-indigenous species are not translocated via ship hulls is to make sure that those ship hulls are free of fouling before they sail. A thorough cleaning of the hull including various niche areas favoured by NIS will avoid such issues.

VOC content of the glass flake vinyl ester STC is less than a third of that of a low VOC content FR coating and less than a fifth of that of a typical AF coating.

3.3 (e) Conclusions STC

The STC approach allows to sail with the most optimal performance as no paint degradation occurs and marine organisms can be removed whenever the shipowner or ship company wants. This reduces the fuel consumption and GHG emissions significantly. Besides, the STC coating offers a lifelong protection and stops the emission of biocides, heavy metals and microplastics into our environment. This approach is therefore 100% environmentally safe and reduces the maintenance and operational costs of the vessel.

4. ECONOMICAL BENEFITS OF AN STC

The economic benefits are best studied in terms of the lifecycle of a ship, comparing the costs and savings associated with different coating types. One such study was conducted in 2006 [4]. The conclusions of the study are expressed in the chart in Figure 6. The numbers, and therefore the variation in costs between the different coatings, would be considerably emphasized if one substitutes today's fuel prices which are significantly higher than those at the time the study was conducted.

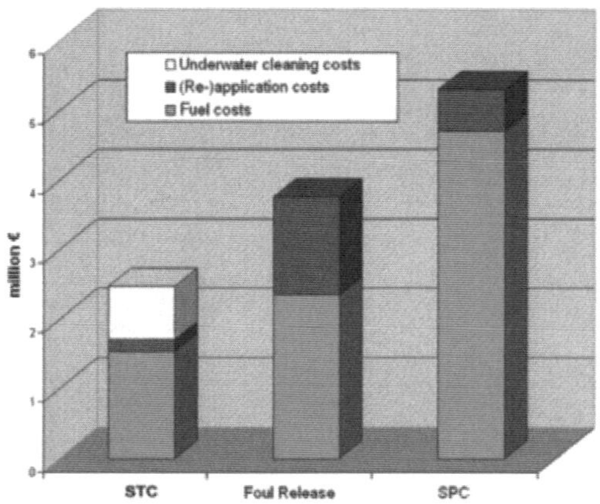

Figure 6 Estimated economic impact (in million €) over 25 years for a 1000-TUE contained vessel with a STC, FR and AF (similar as SPC) [4]

In addition to fuel savings, other advantageous factors which need to be considered in the lifecycle analysis of an STC, are the cost of repeated repainting in dry-dock required by AF and FR coatings, much less time to be spent in dry-docks and the elimination of the costs and environmental damages, such as the release of VOC and toxic materials, associated with traditional recoating. It should be noted that the economic benefits also depend on the marine conditions, the type of vessel and its sailing hours.

5. CONCLUSION

It is easy to simply continue using the conventional methods for the protection of ships' hulls and the control of biofouling which currently prevail broadly. But are these methods, consisting out of the antifouling (AF) and foul release (FR) coating the best available, and are they sustainable? From over 20 years of experience and studies, it was found that these conventional methods do not fulfil the requirements of a coating system.

- The AF and the FR coatings degrade over time which causes a negative effect on the performance of the ship and thus on top speed, manoeuvrability, and long-range capabilities.

- Biofouling still has the change to attach to the ship's hull. Especially when the ship lays idle for a while. This causes for more biofouling and the current systems cannot prevent this. This again results in a reduced performance and unnecessary high fuel consumption and GHG emissions.

- Hull roughness and biofouling can add 25 - 40% or more to a ship's fuel bill. And with a light layer of barnacles or tubeworms, a vessel could see an increase in GHG emissions of up to 55 percent.

- AF and FR coatings are not capable of protecting the vessel against corrosion. These coatings are easily damaged and deplete away over time. The

- The AF coatings and the FR coatings will last for 3-5 years before it must be replaced. This results in a life cycle of re-applicating the coating every drydock session. This involves significantly on material resources, associated transport costs and maintenance costs.

- Moreover, these coating systems have a negative impact on the environmental safety as they release toxic compounds such as heavy metals, microplastics, biocides which end up in enclosed water bodies such as marinas, ports, and harbours. These compounds are proven to be toxic to aquatic organisms, enter the food chain and harm the aquatic environment.

- The VOCs of AF and FR coatings are relatively high. Besides, as these coating systems need to be applied every 3-5 years, the accumulated number of VOCs over the life of the vessel is very high.

- When the vessel remains in port or at anchor for an extended period, fouling is permitted to build up on the AF and FR coating. If they have dug through the coating to the underlying anticorrosive scheme or bare steel, then the likelihood of their washing off is low. Non-indigenous species therefore has the change to be translocated [8].

A better, viable alternative for biofouling control and protection of the ship is the Surface Treated Composite (STC). STCs are formulated for regular in-water cleaning throughout the lifetime of the coating without the need for reapplication. This is an integral part of the hull coating system and can be carried out rapidly at limited cost resulting in improved, rather than repaired or repainted, surface conditions which offer the prospect of significant drag reduction.

- The STC coating does not degrade over time as it is a hard, tough, flexible paint. It is very corrosion resistant as it isolates the steel. This results in the fact that the coating protects the steel of the vessel over its entire lifetime with only minor touch-ups. No repaint required.

- As no corrosion activity is taking place after applying a STC coating, no anodes or other cathodic protection systems are necessary.

- No coating is immune to biofouling. The STC coating will, like the AF coating and FR coating, experience biofouling. However, the STC coating is designed to be cleaned. After cleaning the coating, the fouling will be removed, and the hull will be perfectly smooth as there is no paint degradation over time. Top speed, superior manoeuvrability and long-range capabilities are reached when executed correctly.

- For warships who lay idle for a long time, this approach is much more valuable as the coating can be cleaned before the ship's operational duties. In this way, the ship can sail with the most optimal performance throughout its lifetime.

- The STC coating needs to be applied only once in the condition of sufficient surface preparation. This results in the fact that the coating reaches the lifetime of the ship. This significantly involves reduction on material resources, associated transport costs and maintenance costs.

- STC coatings are 100% non-toxic and inert and therefore does not cause any harm to the environment. In this way the ship sails responsible without any local impact caused by the coating system.

- STCs have the added advantage of offering the potential of eliminating the threat of hull borne invasive aquatic species when the cleanings are carried out correctly.

- VOC content of the glass flake vinyl ester STC is less than a third of that of a low VOC content FR coating and less than a fifth of that of a typical AF coating.

6. ACKNOWLEDGEMENTS

I'd like to thank Boud Van Rompay for his dedication to improve the shipping industry using non-toxic coating systems. The white book "Surface Treated Composites" has provided me with a lot of information for this work. I'd also want to thank David Phillips for his assistance in writing this paper.

7. REFERENCES

1. IMO, "Preliminary results," 2021. [Online]. Available: file:///Users/martijnvanruiten/Downloads/Biofouling%20report.pdf.
2. Svitzer, "Press Release," 26 10 2021. [Online]. Available: https://svitzer.com/below-the-waterline-being-ignored-as-a-prime-lever-in-shippings-drive-to-decarbonise/.
3. Hydrex Group, "Ship Hull Coating Systems Simplified," 2011.
4. E. LIFE, "ECOTEC-STC: Evaluation of a biocide-free hull protection and antifouling system with environmental and economical benefits," EU Life Project, 2006.
5. D. K. D. Y. T. Munk, "The effects of corrosion and fouling on the performance of ocean-going vessels: a naval architectural perspective," In Woodhead Publishing Series in Metals and Surface Engineering, Advances in Marine Antifouling Coatings and Technologies, Woodhead Publishing, 2009.
6. G. Hydrex, "Underwater Ship Hull Cleaning: Cost-effective, Non-toxic Fouling Control," 2011.
7. M. P. J. B. Paola Paruta, "Plastic paints the environment," Environmental Action, Lausanne, 2021.
8. H. Group, "Invasive Aquatic Species," 2011.
9. Hydrex Group, "Hull Coating Degradation – the Hidden Cost," 2012.
10. M. P. Schultz, J. A. Bendick, E. R. Holm and W. M. Hertel, "Economic impact of biofouling on a naval surface ship," Taylor & Francis, USA, 2010.
11. O. T. ,. A. I. S. D. H. I. F. S. D. ,. S. M. O. Y. K. Demirel, "Experimental determination of the roughness functions of marine coatings," 2015.

8. AUTHORS BIOGRAPHY

Martijn van Ruiten joined Subsea Industries six months ago after graduating from the Technical University of Delft with a master's degree in Hydraulic Engineering and a specialization in Ports and Waterways. During his studies, he researched the performance and protection of several coating systems and wrote a thesis on the issue. At Subsea Industries, he can share this expertise with the maritime sector by advising shipowners and shipping companies on how to sail with best performance and lifelong protection while having no environmental impact.

Warship 2022: Designing Warships for the Future
8th-9th of June 2022
M Shed, Bristol, UK

FULL AND SEMI AUTOMATED EARLY-STAGE SHIP LAYOUT GENERATION

N Hifi and **M Courts,** BAE Systems Naval Ships, UK

SUMMARY

The generation of cost-effective ship design solutions to sets of requirements demands a thorough examination of a wide trade space of potential solutions in order to ensure that the ultimately selected solution is the best that can be achieved within an affordable budget. As more options are explored in shorter times then inevitably the degree of detail that can be developed for each solution will decrease, potentially leaving risk that later development will increase the anticipated cost. Early-stage concept design tools are therefore evolving to provide more detail and faster analysis. The importance of ship layout in early-stage ship sizing has long been recognised and methods are required to generate confidence as quickly as possible, and that a practical compartment arrangement can be achieved on any given concept. The paper outlines the general approaches used, and presents example outputs, from a tool developed to make use of genetic algorithms and techniques already proven on existing system generation and analysis toolsets used by BAE Systems Naval Ships. The toolset adopts a staged approach involving a completely flexible mix of automated and manual operations to firstly allocate compartments to zones within a ship and secondly arrange the compartments within those zones. The tool supports the capture and utilisation of knowledge from experienced designers and enables a rapid generation of preliminary ship layout options, within an integrated concept trade space exploration process.

NOMENCLATURE

BAEFASIP	BAE Functional Analysis of System Implementation Parameters
BAEFSM	BAE Force Structure Modeller
BAESAM	BAE System Architecture Modeller
BAESHIP	BAE Ship Sizing and Performance Analysis
BAESLG	BAE Ship Layout Generator
CG	Centre of gravity
GA	Genetic Algorithm
NSGA	Non-dominated Sorting Genetic Algorithm

1. INTRODUCTION

1.1 CONTEXT

Early-stage ship design has often been illustrated by means of a design spiral in which numerous interrelated activities are performed in ever more detail as a solution gradually emerges. The output of the process will include, but not be limited to:

- Specification:
  - Overall dimensions;
  - Weight/space/volume breakdown;
  - Stability assessment;
  - Performance prediction ;
  - Equipment list;
  - General arrangement.
- Cost estimates.

As technology has advanced and warships have become more expensive, all navies are now in the position of having to make compromises in either or both of the capabilities and quantities of ships they acquire and operate in order to meet affordability constraints. Any vessel procurement must therefore ensure that the most cost-effective solution is produced. This is best demonstrated by showing that no better solution is possible at any given budget point, which in turn requires a wide range of potential solutions to be explored, preferably as early as possible in the procurement process. Trade space exploration requires a comprehensive series of activities to be performed involving analysis and trade off of capability requirements against costs. Since achieved capabilities derived from performance predictions, and cost estimates derived from solution specifications, require design solutions to be produced, multiple design spirals must be performed [1]. It is therefore evident that this can only be achieved by carefully automating as many activities as possible, including the traditional design spiral, itself now only part of a much wider design cycle, as illustrated in figure 1.

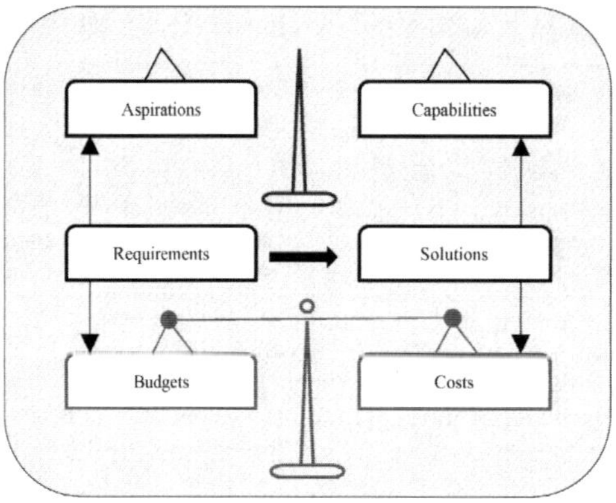

Figure 1: Procurement Balancing Process

The key activities involved in such a process involve [2]:

- **Concept Generation/Engineering Synthesis** - How to generate sufficient reliable data on the full range of viable options of potential interest as quickly as possible?
- **Cost Effectiveness Analysis** – How to bring together cost estimates and disparate operational effectiveness analysis results into a form that can be used to assess the overall value for money of the system;
- **Decision Analysis/Value Assessment** - How to ensure that all possible views and interests are properly accounted for and considered in the analysis and assessment of all the options offered so that an agreed way ahead can be found.

BAE Systems Naval Ships has been developing a tool set to support this process over many years as shown in figure 2. Further tools description and discussion is covered in [2] and [3].

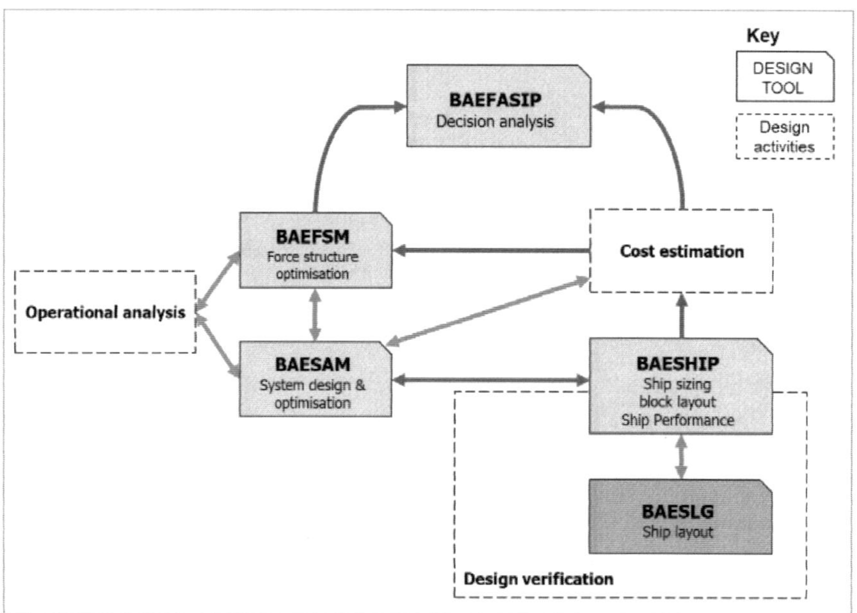

Figure 2: Activities and Design Tools

A key component is the ship sizing and performance analysis tool (BAESHIP), which is used to generate a ship solution to a set of requirements along with sufficient information to generate more detailed performance predictions and cost estimates. Obviously the more design analysis that is incorporated into this tool, then the more confidence can be gained in the options generated. The tool is already extremely capable and includes:

- Weight, CG. area and volume estimation;
- Preliminary tank arrangements and capacities;
- Speed, range predictions;
- Stability analyses;
- Block level layout.

The tool itself can also be programmed by the user to include:

- Crew estimation;
- Cost estimation.

The importance of ship layout in early-stage ship sizing has long been recognised [4] and methods are required to generate confidence as quickly as possible, that a practical compartment arrangement can be achieved for any given concept. The BAESHIP tool has the ability to predefine the general configuration using a parametric block capability together with a facility to check particular areas within the ship against specific space group requirements, using functional blocks. This has proved to be successful in generating ship outline concepts that have subsequently been successfully developed into viable general arrangements. However, the production of even a preliminary general arrangement takes some time even for an experienced naval architect. It has therefore been recognised that it would be desirable to speed up this latter process by the provision of some degree of automation in order to gain more confidence in the solutions generated as part of a trade space exploration study. This would reduce the risk of later development deviating significantly from the initial baseline designs.

The development of such an early general arrangement design tool as part of a concept design system is described by Robb et al [5], Parsons et al [6] and Van Oers [7]. More recent developments in this field includes work by Le Poole et al [8] and Igrec et al [9]. Constructing a general arrangement is an extremely complex process that requires a high degree of knowledge, experience and judgement to be applied. It is evident that the scale of the arrangement problem means that any attempt at automating the process will require:

- The capture of as much expert knowledge as possible;
- A breakdown of the process into manageable steps;
- The ability to apply manual intervention where required.

This paper describes work in BAE that has led to the development of a full and semi-automated ship layout generator tool known as BAESLG (BAE Ship Layout Generator) suitable for initial layout production to a level that gives reasonable confidence that a practical layout can be developed. If the output suggests that a satisfactory layout is not achievable, then there is enough information to indicate where overall ship envelope changes are required.

1.2 OBJECTIVES

The objectives of the tool are to assist the designer developing a concept to:

- Understand the major layout drivers;
- Increase confidence that a practical layout can be achieved;
- Speed up initial layout development;
- Provide a reliable start point for baseline development.

It was also recognised that any tool must be both flexible and scalable to cope with a wide range of ship types.

2. PROBLEM CHARACTERISTICS

The problem of developing a satisfactory ship compartment layout is an architectural arrangement task with the following characteristics:

- Competing requirements for compartment positions;
- Effect on operational efficiency;
- Convenient access to compartments;
- Adequate means of escape from compartments;
- Relative position of compartments to system routes.

All to be achieved in the limited space available. Success in achieving a satisfactory layout can be measured by defining factors to determine how well the requirements are met, Robb et al [5]. This leads to the formulation of an overall optimisation problem characterised by:

- Multiple objectives;
- Huge number of possible solutions;
- Uneven solution space with many local optima.

3. SELECTED APPROACH

In order to make the problem manageable, it has been broken down into multiple allocation stages:

- Allocation of compartments to major zones of the ship, typically but not limited to, regions between bulkheads;
- For each zone:
    - Allocate compartments to sub zones within each zone, typically but not limited to, individual decks;
    - For the compartments in each sub zone:
        - Generate a compartment layout by allocating compartments to defined positions in the sub zone and growing each compartment around its allocated position.

At each stage, manual intervention is possible to either:

- Replace the automatic operation with a manual allocation;
- Modify the automatic allocations.

If required, the final layout can also be manually edited and improved using an interactive graphical editor.

A further simplification of the allocation problem is achieved in practice, because some major compartments have already been positioned during the initial sizing and block layout generation process conducted by BAESHIP. Typically, the compartments already positioned will include:

- Main machinery rooms and uptakes;
- Steering gear compartments;
- Chain lockers etc.;
- Hangars;
- Main magazines;
- Vertical launch missile systems;
- Mission bays.

The final compartment layout stage can be simplified by predefining some or all of the major access routes, both passage ways and vertical access, through the ship. These access routes are likely to house many of the major system distribution routes. The location of these routes is often a design decision influenced by a variety of factors:

- Operability;
- Vulnerability;
- Evacuation;
- Producibility.

Each of the allocation stages listed at the start of this section can be considered as a separate operation that can be optimised against defined objective parameters. It can be appreciated that since the assessment of the suitability of any given optimisation is, among other factors, dependent on the relative position of one compartment to another, then the overall result of the optimisations will depend on the order in which they are conducted and whether the cycle is repeated or not. In practice, any evaluation order effects are likely to be secondary if the selection of zones and sub zones is conducted carefully, with due regard to the zones' practical implications on ship layout.
In order to provide as much flexibility in the tool as possible a combination of approaches are provided:

- **Completely manual:** All compartment allocations to zones, sub zones and positions can be performed manually with the tool providing feedback on the areas available and used and scores against requirements;
- **Fully automated:** The entire process can be performed completely automatically with the optimisations, selection of preferred solutions and compartment layout all being performed by in built algorithms;

- **Mixed auto/manual:** A mix of automatic and manual operations is possible. The result of each automatic operation can be manually edited before the next step in the process, or performed completely manually.

4. LAYOUT ASSESSMENT

The success or otherwise of each stage of the compartment allocation and layout process requires measurable parameters that can be assessed against specified requirements. The initial input data comprises a list of compartments. This list will usually not include minor local storage spaces, etc. For each compartment specified the following data is required:

- Size and area;
- Absolute position range in the ship specified by longitudinal position and relative position to the deck and superstructure;
- Access position;
- For multiple instances of the same type of compartment:
  - Spreading throughout ship;
  - Clustering together.
- Objectives for relative position of compartment to other compartments, including blocks already defined in BAESHIP;
- Compatibility with removal routes.

All the above requirements are assessed in the calculation of benefit parameters. Assessment against these parameters results in a series of scores, which as far as possible decrease linearly the more the requirement is infringed. In this way optimisers are assisted in finding the best solutions and the results of any manual intervention can be fed back to the user. Obviously the degree to which all the above requirements are satisfied and scored will depend on the allocation stage under consideration and the associated degree of compartment definition available.

5. ZONE ALLOCATIONS AND OPTIMISATION

In order to perform the automatic compartment to zone or compartment to sub zone allocations, optimisation objectives are required. Any calculated parameter within the software can be optimised or constrained but the default parameters are set as:

- Maximise overall total compartment score calculated using all available requirement assessment parameters for all compartments (in range 0.0 to 1.0);
- Maximise overall minimum void space ratio for the zones or sub zones being considered. This has the effect of spreading the compartments out across the available space (in range 0.0 to 1.0).

In order to ensure that sufficient space is available for subsequent layout development a minimum access ratio can be specified for the zones or sub zones. The void space used in the objective is then the deck area available minus the compartment areas and any defined access areas.

6. COMPARTMENT LAYOUTS AND OPTIMISATION

A simple method is used to generate the compartment layout within a sub-zone. The method uses a rectangular grid with each resultant cell being allocated to one of:

- Available deck area;
- BAESHIP defined block;
- Access route;
- Compartment.

The grid dimensions can be set to be sub-divisions of a frame spacing and can differ in the x and y directions. The available cells on a given deck are set up to be within deck edges and allow for any tumblehome in the hull or superstructure. Access routes including passage ways and vertical access can be assigned to cells manually using a graphical editor if required. This can be done once the compartments assigned to the sub-zone are known from the appropriate compartment to sub-zone allocation optimisation, which can itself be performed assuming a default access ratio.

Once the cells and any access routes have been set up an array of nodes is defined. Nodes can be defined in one or both of two ways:

- Along passage ways thus ensuring all attached compartments have appropriate access to a passageway;

- Regular grid across whole sub zone. Any attached compartment may not have access to a user defined passageway.

Compartments are then assigned to nodes and given a length, either by the optimiser or by the user for a manual allocation. A compartment growing algorithm then grows the compartment as best it can from the assigned node assigning each enclosed cell to that compartment. As compartments are grown one after the other it could be expected that priority would be given to the first compartments in the list. In fact the optimiser's search for suitable nodes tends to reduce any priority effect. In addition, the growth algorithm seeks to compensate for the node not being centrally placed in the compartment, thus further reducing any priority effect and speeding up the optimisation process itself.

If nodes that are not attached to existing access routes, i.e. grid nodes, have been used, then a passage growth algorithm is implemented which grows passages out from the main access routes and thus connects compartments not on the access routes where possible. Two stage growth is used for the passages, i.e. initially at right angles from the main routes and then again at right angles to the new passages.

Once the compartments and sub passages have been completed the cell arrangement is assessed against all the requirements and appropriate scoring parameters calculated. For the automatic layout optimisation process the following optimisation objectives are defined:

- Minimise total compartment area lacking. This will be greater than zero if the growth algorithms were unable to achieve the required compartment area;
- Maximise overall total compartment score calculated using all available requirement assessment parameters for all compartments (in range 0.0 to 1.0);
- Maximise total number of acceptable compartments. A compartment is defined to be acceptable if:
  - The required area is met;
  - The length and breadth are greater or equal to the required minimum values;
  - It is attached to a passage.

7. GENETIC ALGORITHM

The optimisation method used for all automatic allocations in the software is a Genetic Algorithm (GA) based on the Non Dominated Sorting GA II/III (NSGA), [10] and [11]. It has been augmented with:

- Local optimisers designed to improve the results for a specified proportion of each generation's population members using a choice of simple methods;
- Constraint functions;
- Requirement compliance enforcers applied to a proportion of population members.

The optimiser is designed to generate results along the Pareto front (surface) of optimised solutions. If the spreading function is turned off then the results will tend to cluster together.

The compartment to zone and sub-zone optimisations are generally run as a 2D optimisation as described above. The selected solution is picked from the front either manually or by using an algorithm that picks the nearest solution to a line through the ideal solution (1.0, 1.0) with a specified slope and with a minimum void area ratio specified. Whether the crowding function is used or not is a user choice and will be dependent on the choice of slope used in the automatic solution selection or the range of options desired for a manual solution selection. Example results from a compartment to zone optimisation for a Multi Role Ship are shown in Figure 3 and Figure 4.

The compartment to node allocations for a sub-zone are generally run with the crowding function activated so that as wide a spread of solutions as possible are generated. The automatic selection algorithm will then pick the solution with the maximum number of compartments that are appropriate and with the least area lacking. This selection can be overruled by the user. Results from a typical compartment to node optimisation are shown in Figure 5 illustrating the compromises possible. Each point represents a possible layout solution for the sub-zone involved.

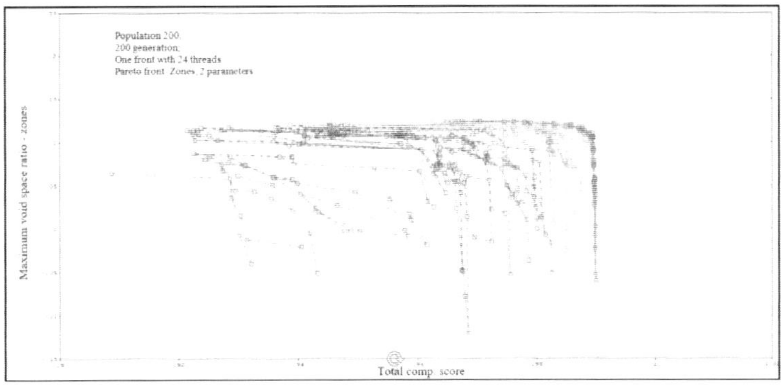

Figure 3 Typical GA Optimum Front Generation History of All Compartments to Zone Optimisation

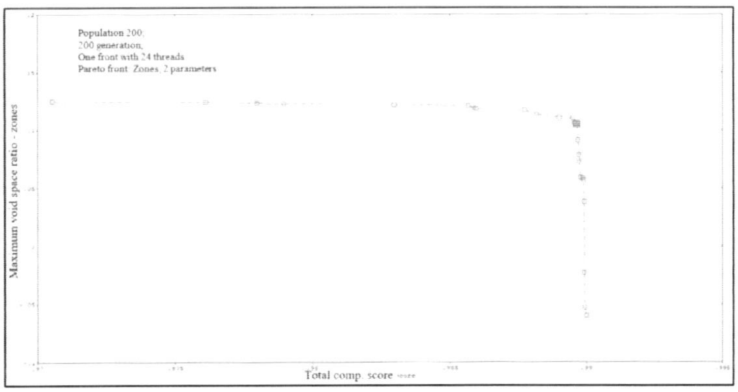

Figure 4 Typical All Compartments to Zone Optimum Pareto Front Result and Auto Selected Solution

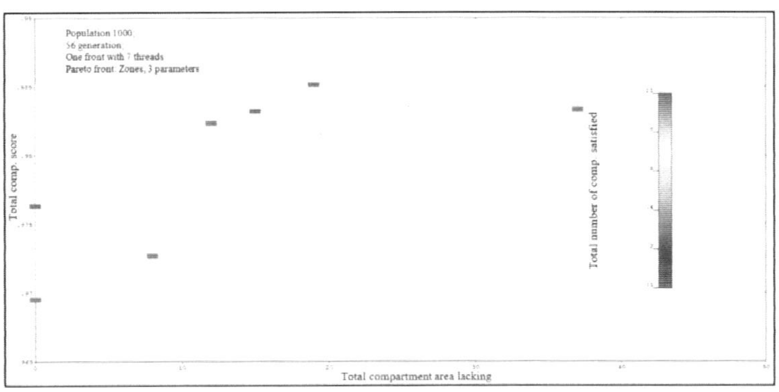

Figure 5 Typical Compartment to Node Optimum Pareto Front Result for a Sub-Zone

8. EXAMPLE RESULTS

Example results presented in this section are for two deck layouts resulting from a fully automatic sequence. The first one shown in Figure 6 is one of the deck allocation results for an Offshore Patrol Vessel with a ship total of 105 individual compartments defined. The second result in Figure 7 is for one of the decks for a Multi Role Ship with a ship totalling 266 compartments. In both cases a simple arrangement of main access routes was predefined at the start of the process by the user and no attempt was made to either update the routes or alter the outputs from each allocation stage before the next was conducted.

The layouts show the zones and sub-zones used in the allocation stages together with one deck layout. Any issues to be resolved are highlighted by the compartment colour coding. Individual compartments can be identified either by placing the cursor over them or zooming the display.

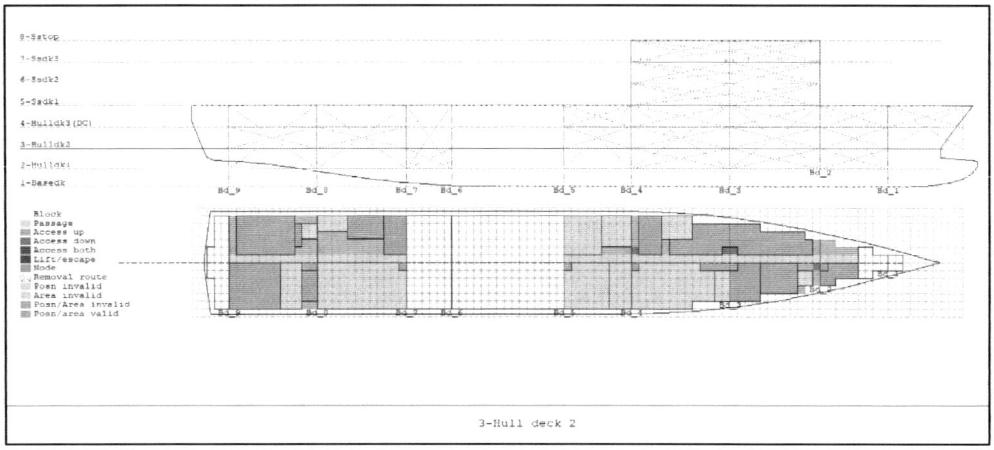

Figure 6 Example Offshore Patrol Vessel Zones and Deck Layout from Sequence of Automatic Allocations

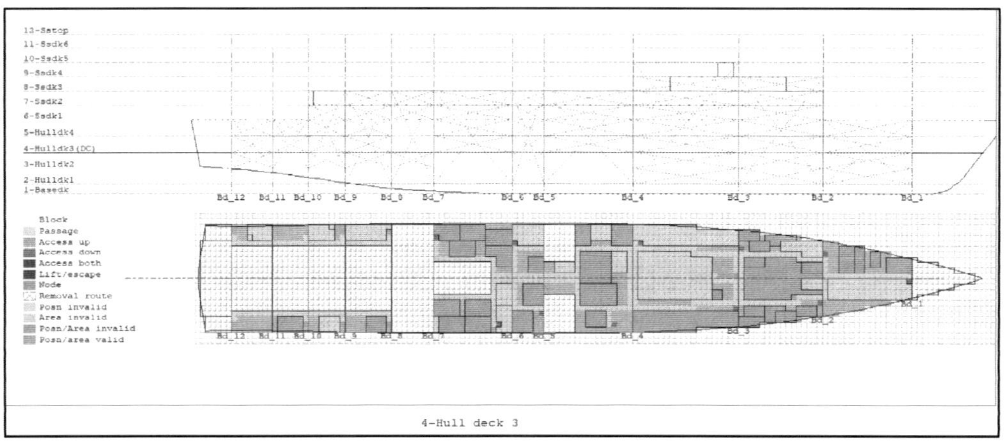

Figure 7 Example Multi Role Ship Zones and Deck Layout from Sequence of Automatic Allocations

In both cases, the plots show that there is spare area available. It can also be seen that various issues exist for some compartments:

- Compartments are not in the correct position in the ship;
- Compartments are not in the correct position relative to all others;
- Compartments cannot grow to sufficient area or satisfactory shape because of the position and shape of other compartments and the positions chosen for the major access routes and indeed bulkheads themselves;
- Compartments are simply not accessible.

Detailed information for each compartment is available to the user and can be displayed by placing the cursor on a compartment. This can assist the user if the in-built graphical editor is used to improve the layouts.

9. DISCUSSION

All the allocation optimisations inevitably involve compromises at many levels:

- Trade-off between selected optimisation variables;
- Implied weightings on requirements inherent in the calculation of result parameters. Since requirements are quite likely to conflict some will not be met in any given solution.

For the optimisation variable trade off the user can choose from a selection of results on a Pareto front of optimised solutions. However, the second compromise is far more problematic for several reasons:

- The software caters for a limited set of possible compartment requirements. These have been selected to cover as wide a range of possibilities as possible but cannot cater for all possible scenarios. They can of course be added to, if necessary, in future software upgrades;
- The allocation results will depend heavily on the user supplied requirements. This requires input from experienced designers and will inevitably reflect their particular preferences;
- No attempt is made to weight the different requirement options. Although this is entirely feasible it is felt that in practice it would confer very little benefit on the overall process;
- The more requirements are defined then the less effect each requirement has on a single aggregated overall benefit optimisation parameter.

The net effect is that no fully automated process will produce completely satisfactory results and a degree of manual intervention will always be necessary at each stage to produce an acceptable layout. This will probably extend to bulkhead position and zone adjustment and maybe changes to the overall layout concept requiring iteration with BAESHIP. The software includes an option that allows the user to edit and manually update the zone and sub-zone allocation solutions, which provides instant feedback on the effects of any changes made, see Figure 8.
It is of course not necessary to define all compartments in detail and, particularly for larger vessels, aggregation of compartments into functional areas is desirable.

The balance between numbers of compartments or functional areas and size of ship, together with the optimiser settings selected will govern both the result quality and execution time of the allocation optimisations. Depending on optimiser settings the total execution times for the ship examples presented in fully automatic mode can vary between 10 minutes to 2 to 3 hours. Although the auto selected solution is presented in the example layout each zone may have multiple options on its allocation front making a large number of potential layouts available to the user on a pick and mix basis.

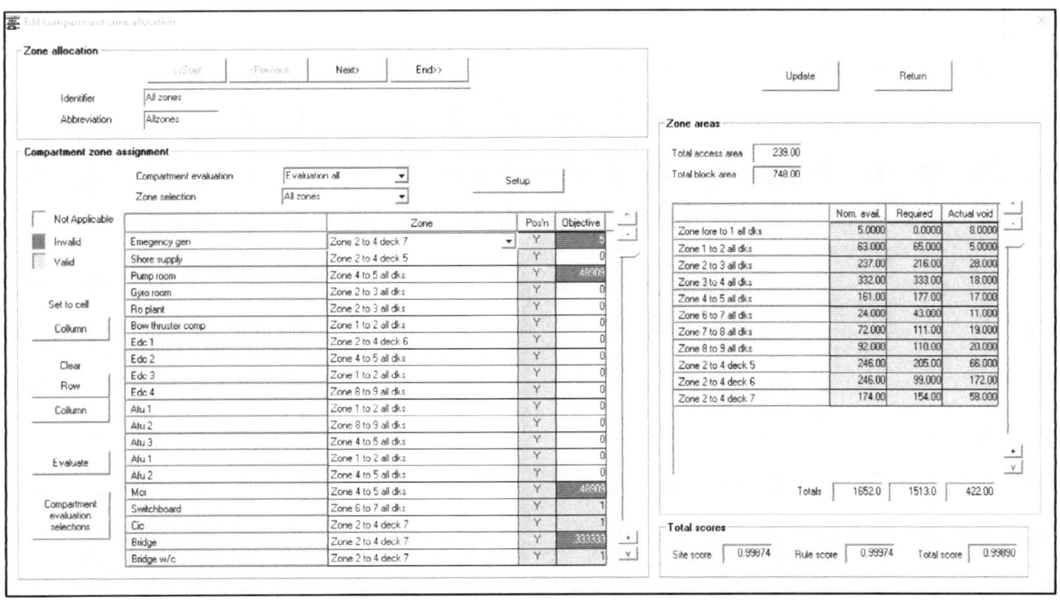

Figure 8 Manual Editing of Compartment to Zone Allocation

10. CONCLUSIONS

Any layout generation process is heavily dependent on expert input; Even if the fully automatic options are used then the results are very dependent on the input data provided by experts, particularly the compartment requirements. The breakdown of the process into logical steps makes the processes far more manageable and allows the optimisers to perform better on what is in theory an impossibly large trade space of potential solutions. Although the tool allows both completely manual and fully automated operations the best results are probably achieved by using the automated allocations in conjunction with manual editing.
In practice, the user can often appreciate whether a concept is likely to be viable without having to go through the entire process. Although still evolving, the tool has demonstrated that it can fulfil its original objective to provide confidence that a practical layout is achievable beyond what can be gained from simple area analysis. The automatic processes have been shown to achieve this by quickly illustrating where problems occur. It is always easier to see where the real solutions lie when the problems in potential solutions are made visible.

11. ACKNOWLEDGEMENTS

The authors thank BAE Systems Naval Ships for permission to publish this paper and would like to acknowledge the help of their colleagues in its preparation.

12. DISCLAIMER

The views expressed in this paper are those of the authors and do not necessarily represent those of BAE Systems Naval Ships.

13. REFERENCES

1. M. Courts, B. Durant and T. M, "Affordable Warships – Understanding the possible," in *The Affordable Warship*, Bath, UK, 2007.
2. M. Courts, R. Brittain, A. Lamble and N. Osborne, "Warship Trade Space Exploration: Challenges and Approaches," in *RINA Systems Engineering Conference*, London, 2012.
3. M. Courts, "The Use Of Functional Analysis To Ensure Value For Money In Complex Engineering Projects," in *International Conference on Computer Applications in Shipbuilding*, The Netherlands, 2019.
4. D. Andews, "The Sophistication of Early Stage Design for Complex Vessels," Transactions RINA, Special Edition, IJME, 2018.
5. M. Robb, J. Mykura, N. Healey , P. Rottier and S. Bryce, "Design of Efficient General Arrangements: a software aid to the decision making process," INEC Conference, Glasgow, 2002.
6. M. Parsons, K. E., H. Chung and S. DANIELS, "Intelligent Ship Arrangements: A New Approach to General Arrangement'," Naval Engineers Journal, 2008.
7. B. Van Oers, "A Packing Approach for the Early Stage Design of Service Vessels," TU Delft, 2011.
8. J. J. Le Poole, ,. E. A. E. Duchateuau E. A. E, B. J. Van Oers and A. A. Kana, "A Case Study Into An Automated Detailed Layout Generation Approach in Early Stage Naval SHIP Design," 15th INEC Conference, 2020.
9. B. Igrec, R. Pawling, G. Thomas, A. Sobey and J. Rigby, "An Interactive Layout Exploration and Optimisation Method for Early Stage Ship Design," ICCAS, Rotterdam Netherlands, 2019.
10. K. Deb and H. Jain, "An Evolutionary Many Objective Optimization Algorithm Using Reference Point Based Nondominated Sorting Approach, Part1: Solving Problems with Box Constraints," IEEE Transactions on Evolutionary Computation, Vol. 18, No. 4, August 2014.
11. K. Deb, A. Pratap, S. Agarwal and T. Meyarivan, "A Fast and Elitist Multi-objective Genetic Algorithm: NSGA-II," IEEE Transactions on Evolutionary Computation, Vol. 6, No. 2, April 2002.

14. AUTHORS BIOGRAPHIES

Nabile Hifi is a Senior Technologist in BAE Systems specialising in design for automation, modelling and analysis.
Malcolm Courts is a Consultant to BAE Systems with involvement in concept development and design process improvement.

DESIGN FOR SUPPORT IN INITIAL DESIGN USING THE NETWORK BLOCK APPROACH: A SUBMARINE EXAMPLE

M H Mukti, R J Pawling, and **D J Andrews,** University College London, UK

SUMMARY

The layout of the Distributed Ship Service Systems (DS3) can be a major constraint both on the location of hull berth unit and of transverse bulkheads, with implications for conducting maintenance and so contribute to submarine Sustainability. This paper proposes a new early-stage ship design (ESSD) approach that would aid the designer to consider DS3 features physically and logically much earlier in the design process. The approach is termed the Network Block Approach and combines the advantages of the UCL 3D physically based ship synthesis Design Building Block (DBB) approach and the network-based approach for distributed ship service systems (DS3) synthesis already proposed for naval combatants. The proposed Network Block Approach enables modelling of individual spaces to locate and zone various DS3 components and take account of hull unit boundaries and major bulkheads. The paper discusses the importance of DS3 routing to warship Sustainability through the complexity of a submarine design example and shows how the proposed approach enables this to be explored in ESSD.

NOMENCLATURE

AFO	Architecture Flow Optimisation
AIP	Air Independent Propulsion
ATP	Air Turbine Pump
CASD	Computer Aided Ship Design
ESD	Early Stage Design
ESSD	Early-Stage Ship Design
DBB	Design Building Block (approach)
DS3	Distributed Ship Service Systems
FMFI	Fight Move Float Infrastructure (functional breakdown)
MIT	Massachusetts Institute of Technology
NBA	Network Block Approach
NFO	Network Flow Optimisation
S^5	Speed, Strength, Stability, Seakeeping, Style (naval architectural concerns in vessel design)
UCL	University College London

1. INTRODUCTION

Submarine design poses significant problems related to the complexity of submarine Distributed Ship Service System(s) (DS3) (Andrews, 2017a) since these systems must be physically arranged in an extremely limited space while ensuring their vital functioning can be maintained through Sustainability consideration. Sustainability is significantly affected by the architecture of the vessel by providing appropriately located spatial margins for future equipment and provision ease of access to and removal of equipment to assist maintenance (Andrews & Pawling, 2003). However, this has traditionally been neglected since the DS3 physical models were considered to take too long to define sufficiently in ESSD (Rigterink, 2014). This paper shows how improved modelling of DS3 in early-stage design can be used to make better decisions on the submarine's Sustainability and provide adequate overall size and layout for a conventional submarine design and thus reduce the need for expensive rework in later design phases.

The paper begins with a framework to cover design style decisions that need to be made at ESSD. This is followed by the description of a new computer-based Network Block Approach making the interrelated process of network-based distributed system synthesis and 3D geometric submarine design much more effective and time-efficient, while retaining designer control over key decisions. A case study to validate the capability of the approach is then outlined. Finally, conclusions are drawn, and ongoing work is presented.

2. STYLE FRAMEWORK

In the initial sizing of complex vessels, there are front-end decisions that are traditionally implicit but have been argued should be included if Requirement Elucidation is to be adequately undertaken (Andrews, 2018). These decisions are very important because they will drive the final chosen design solutions (Andrews, 2018). They can also influence the ship design 'traditional naval architecture' S^4 primary concerns (Brown & Andrews, 1980) namely those of Speed; Strength; Stability; and Seakeeping (although manoeuvrability is clearly a more appropriate concern for the submarine design than that of Seakeeping). Brown & Andrews (1980) saw that there were wider largely whole ship issues and introduced the term 'style' to add a fifth S.

A framework for style decisions can be considered as shown in Figure 1 to cover four different levels of style decisions. Those levels of style choice can be considered from high-level decisions to more detailed concerns. The highest level is that of mega style, which largely reflects the government selection (Andrews, 2021). An example of high-level endorsed style decisions at the *Macro Level* would be to adopt nuclear or non-nuclear for a new submarine programme. Mid-level could address *Main/Meso Level* style choices, which included decisions for technology adopted in a submarine design, such as various AIP types. A more detailed style choice, seen to be at *Micro Level*, and appropriate to adopt specific DS3.

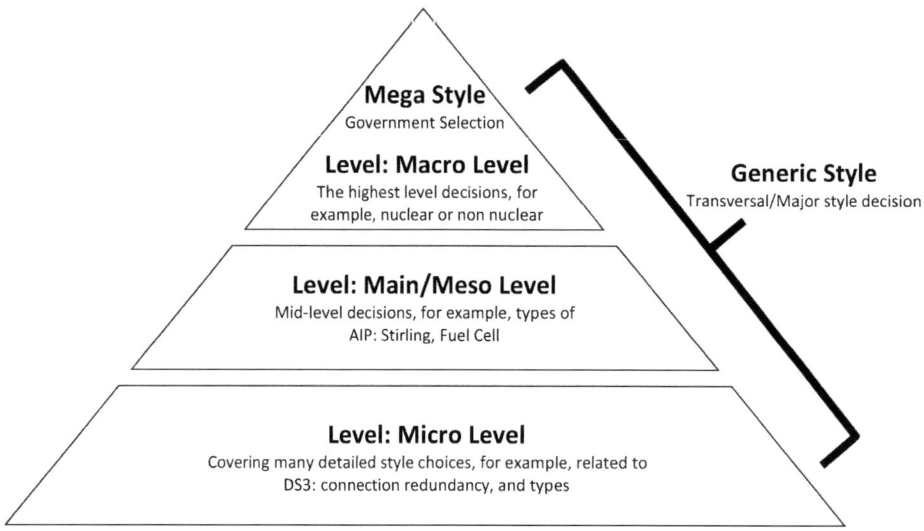

Figure 1. Style framework.

Style categorisation can reflect the transversal style decisions (Andrews, 2017b) as listed for a SSK in Table 1. This generic level of style will influence either directly or indirectly other levels of style choice. Therefore, for example, if a generic style choice, such as robustness, was to be selected, the realisation of that robustness style choice at micro-level style could be a ring main DS3 configuration, which would give a high level of redundancy (port and starboard, forward and aft connections) and thus robustness to the DS3 design. Andrews (2018) considers these decisions need to be consciously undertaken in ESSD requiring to be acknowledged by the designer and justified and then documented as a mark of good design practice in the decision making process for complex vessels.

Table 1. Listing of style topics relevant to submarine design (Andrews, 2021).

Stealth	Protection	Human Factors	Sustainability	Margins	Design style
Acoustic signature	Underwater weapon effect	Accommodation & Escape	Mission duration	Space	Robustness
Magnetic	Fire	Access	Watches	Weight	Adaptability
Infra-red	Shock	Maintenance levels	Stores	Vertical & Longitudinal centre of gravity	Modularity
Radar cross section	Damage control	Operation automation	Maintenance cycles	Power	Operational serviceability
Visual	Collision	Ergonomics	Refit philosophy	Services	Producibility
	Above water Weapon effect		Upkeep by exchange	Board Margin (future upgrades)	
	Corrosion				

The framework for various style decisions has been outlined. The following sections present how the proposed approach for DS3 could facilitate this.

3. THE NETWORK BLOCK APPROACH

The proposed computer-based approach, referred as to the Network Block Approach (NBA) (Mukti, 2022), was devised to consider networked DS3 in the early stage design process of submarines. Traditionally left to later design phases, the proposed approach aims to demonstrate that it is possible and important to look at piping, cabling, and HVAC networks during Early-Stage Design (ESD) before a submarine's size and dimensions are fully determined. The proposed approach employed a network-based synthesis tool called SUBFLOW (Mukti et al., 2021), a series of spreadsheet-based programs (Mukti et al., 2022a), and adapting the existing CASD tool Paramarine-SURFCON (Qinetiq, 2019) used by the ship design research team at UCL (Mukti et al., 2022b), to make the inter-related processes of network and overall physical design much more efficient.

Thus, the Network Block Approach utilised DevOps software practice which blends 'Development' and 'Operations' activities (Hüttermann, 2012). This leads to an infinity loop diagram presented in Figure 2, which represents the iterative nature of the process. This process can be terminated once the design is considered naval architecturally balanced. Figure 2 shows (centrally) the Excel spreadsheet (Microsoft, 2021), where the design data can be stored and is termed the 'input data centre'. The design data deals with the design concerns appropriate to ESSD, such as weight and space. The NBA consisted of the 'Physical Loop' method shown in purple in Figure 2 (left) and the 'Logical Loop' method shown in blue in Figure 2 (right).

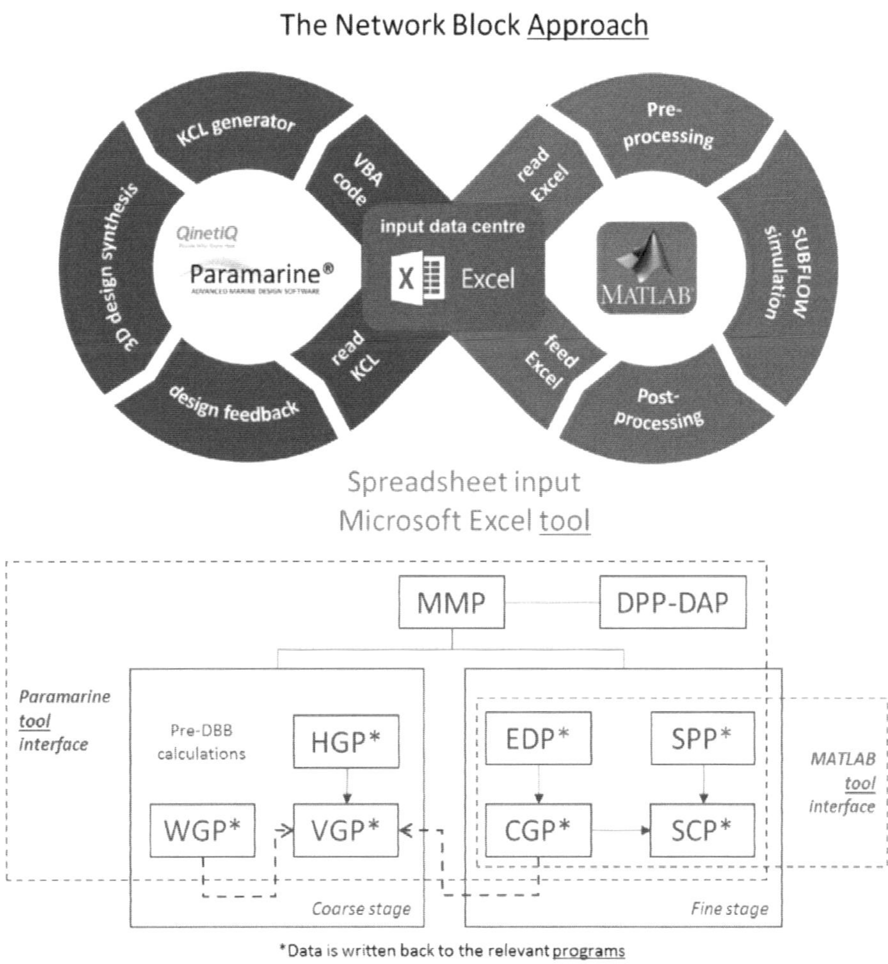

Figure 2. Summary of the proposed Network Block Approach (NBA) consisting of proposed methods (Physical Loop method in purple, inputs in green – see Table 2 for acronyms of the programs, and Logical Loop method in blue).

The Physical Loop method focuses on synthesising the submarine design and the DS3 in terms of the physical architecture. This utilised the 3D based synthesis UCL DBB approach (Andrews et al., 1996) that is readily facilitated in Paramarine-

SURFCON software (Pawling & Andrews, 2011). The Logical Loop method makes use of MATLAB codes (MATLAB, 2019) and CPLEX (IBM, 2014) to perform the development of a DS3 network by pre-processing, analysis, and post-processing through the new tool (SUBFLOW) to enable DS3 energy flow simulation representing the logical architecture (Brefort et al., 2018) and an energy based DS3 sizing. SUBFLOW is a network-based approach that was derived from the MIT Network Flow Optimisation (NFO) approach (Trapp, 2015) and the Virginia Tech Architecture Flow Optimisation (AFO) approach (Brown, 2020). There were two possible ways to synthesise DS3 using SUBFLOW:

- Sketch and develop the SUBFLOW network of a DS3 configuration making specific style choices and then derive the length to weight and volume ratios for the relevant DS3.

- Break down the power demands of each DS3 component in the SUBFLOW network to undertake an early-stage steady-state flow simulation. This numerical energy flow information can then be translated into DS3 space and weight inputs by deriving power to weight and power to volume ratios.

The first way is quick, less effort but is not sensitive for sizing DS3 that are network flow-dependent, such as the heat removal system. The latter can be used to ensure an early-stage energy balance but it is more demanding as it requires more engineering definition and inputs. In the Network Block Approach, the DS3 SUBFLOW network styles were proposed on the basis of prior expert knowledge and were deliberately not validated by analysis in ESSD. This not only allowed a more style-driven approach but also allowed the SUBFLOW process to be as efficient as possible, commensurate with the needs of Requirement Elucidation (Andrews, 2011) in submarine ESD. The SUBFLOW network can nonetheless provide a suitable basis for modelling (say) survivability concerns in subsequent design phases, if required.

The input data centre (Figure 2) consisted of several programs, which are listed in Figure 2 (bottom) and detailed in Table 2. Each program was developed as a worksheet, which could be read both: by Paramarine for rapid modelling of objects; and by MATLAB for automatically generating multiple setups to perform SUBFLOW. With this approach, the designer could focus and readily manipulate the architecture of the overall vessel and perform SUBFLOW analyses without needing to address the demanding modelling setup in Paramarine-SURFCON and MATLAB.

Table 2. Summary of programs in the Network Block Approach.

Program	Description	Function
MMP	Main Menu Program	Execution menu to compile all programs
DPP	Design Preamble Program	Hardcoded design setup
DAP	Design Analysis Program	Hardcoded analysis setup
HGP	Hull Granularity Program	Input for hull size
VGP	Volume Granularity Program	Input for spaces
WGP	Weight Granularity Program	Input for weight
EDP	Equipment Database Program	Input for equipment data
CGP	Component Granularity Program	Input for DS3 components for arrangement and SUBFLOW
SPP	System Preamble Program	Input for DS3 connections
SCP	System Connection Program	Input for DS3 connection and SUBFLOW

With the Network Block Approach briefly described, the next section presents a case study showing it is possible to carry out meaningful modelling of such networked systems in early design.

4. SUBMARINE CASE STUDY

To show how important DS3 are in Sustainability and how the proposed approach enables this to be explored in ESSD, a generic submarine case study was developed using the submarine database used in the annual UCL submarine design exercise (UCL-NAME, 2014). This is spelt out for the SSK example in Table 3.

Table 3. The realisation for the presented submarine case study, which follows the decision making sequence for complex vessels outlined in Figure 4 and Appendix A of Andrews (2018) in a similar manner to the submarine example in Figure 4 of Andrews (2021).

<table>
<tr><th>Process Step</th><th colspan="3">Selection Decision / Realisation for the Case Study</th></tr>
<tr><td>Perceived need</td><td colspan="3">Demonstrate the application of the Network Block Approach (NBA) for Sustainability</td></tr>
<tr><td rowspan="16">Outline of initial requirements</td><td colspan="3">Initial submarine key parameters desired (capabilities)</td></tr>
<tr><td>Payload equipment</td><td colspan="2">Description</td></tr>
<tr><td>6 x 600 mm tubes (OD)</td><td colspan="2">Torpedo tubes to eject the torpedo from the submarine</td></tr>
<tr><td>18 x Spearfish-UK</td><td colspan="2">Torpedo reloads as the anti-ship or anti-submarine weapon</td></tr>
<tr><td>2 x Air turbines</td><td colspan="2">Air turbine pump torpedo discharge system</td></tr>
<tr><td>2 x Bandfish-UK</td><td colspan="2">Countermeasures to distract, decoy or destroy incoming torpedoes</td></tr>
<tr><td>SPA4 RFOM-81</td><td colspan="2">Sonar systems (1 cylindrical Bow Passive Sonar (BPS), 6 Passive Ranging Sonar (PRS), and 2 Flank Array Sonar (FAS))</td></tr>
<tr><td>R3 RFOM-100</td><td colspan="2">Radar mast non-penetrating telescopic</td></tr>
<tr><td>EW4 RFOM-72</td><td colspan="2">Submarine Electronic Warfare</td></tr>
<tr><td>C5 RFOM-53</td><td colspan="2">Communications fit, which operates across the frequency spectrum</td></tr>
<tr><td>CMS7 RFOM-85</td><td colspan="2">Combat Management Systems</td></tr>
<tr><td>Initial performance</td><td colspan="2">Value</td></tr>
<tr><td>Complement</td><td colspan="2">46 personnel</td></tr>
<tr><td>Operational environment density range</td><td colspan="2">1-1.025 te/m^3</td></tr>
<tr><td>Deep diving depth</td><td colspan="2">250 m</td></tr>
<tr><td>Patrol length</td><td colspan="2">49 days</td></tr>
<tr><td rowspan="7">Selection of Style of emerging ship design</td><td>Style level</td><td colspan="2">Choice</td></tr>
<tr><td>Generic level</td><td>Sustainability</td><td>See Table 1</td></tr>
<tr><td rowspan="2">Macro level</td><td colspan="2">Medium size ocean-going submarine</td></tr>
<tr><td colspan="2">Single hull with casing</td></tr>
<tr><td rowspan="2">Main level</td><td colspan="2">Conventional (diesel-electric) power plant</td></tr>
<tr><td colspan="2">Two watertight bulkheads</td></tr>
<tr><td>Micro level</td><td colspan="2">Various detailed DS3 styles</td></tr>
<tr><td>Selection of major equipment and operational sub-systems</td><td colspan="3">DS3 configurations (see following subsections and Chapter 5 in Mukti, 2022)</td></tr>
<tr><td rowspan="7">Selection of whole ship performance characteristics</td><td>General performance</td><td colspan="2">Value</td></tr>
<tr><td>Sprint speed</td><td colspan="2">20 knots (2 hrs)</td></tr>
<tr><td>Submerged speed</td><td colspan="2">5 knots (17 hrs)</td></tr>
<tr><td>Snort speed</td><td colspan="2">6.5 knots (5 hrs)</td></tr>
<tr><td>Indiscretion ratio</td><td colspan="2">22%</td></tr>
<tr><td>Sub level</td><td colspan="2">Value</td></tr>
<tr><td>For DS3</td><td colspan="2">See Section 4</td></tr>
<tr><td>Selection of synthesis model type</td><td colspan="3">The Network Block Approach (NBA) (Mukti, 2022)</td></tr>
</table>

Having outlined the decision making process for the case study, the next section describes the development of the case study *ab initio*, which includes the application of Physical Loop and Logical Loop methods (see Figure 2).

4.1 A ZONAL APPROACH TO DS3 DESIGN

As specified in Table 3, the case study was deliberately developed with design for support in mind as a selected generic style choice for the case study as part of a Requirement Elucidation approach to ESSD (Andrews, 2018). To significantly improve the level of access to conduct maintenance and indeed to enable the movement of equipment to a different environment for the purposes of maintenance, the use of a clear opening (e.g., 2.2 m x 1.9 m) referred as a shipping opening, or "Dutch breach" has been a feature of French submarines for years and was originally adopted by the Royal Netherlands Navy and then in the UK UPHOLDER Class (Wrobel, 1984). A further approach was to undertake a full circumferential hull cut (see Figure 3), as already implemented by the Australian Navy for its COLLINS Class to refurbish motor and generators (Morris & Spinney, 2017). This allowed the simultaneous overhauling process of those components in the workshops and the residual components in the compartment, contributing to submarine's Sustainability (Coles, 2014). Thus, such a full circumferential hull cut approach was adopted when developing the concept design and investigated in the current case study.

Figure 3. Collins-class submarine full-cycle dockings are complex activities in which the back end of the submarine is cut off to allow the motor and generators removal (Hellyer, 2018).

To implement such a generic style choice, the case study was designed with four maintenance sections, allowing access to major DS3 components (e.g., propulsion motor, diesel engines, aft and forward battery groups):

- Aft maintenance section;
- Machinery maintenance section;
- Mid maintenance section; and
- Forward maintenance section.

Therefore, the submarine synthesis was commenced by undertaking the traditional numerical synthesis using a simple synthesis spreadsheet, estimating the initial crude weight and space balance of the design in the manner of a pre-DBB synthesis. Such a traditional numerical synthesis relatively helped the designer to size the submarine *ab initio*. Subsequently, the Weight Granularity Program (WGP) and the Volume Granularity Program (VGP) could have been used to architecturally manipulate the arrangement of the submarine at the coarse stage (Figure 4).

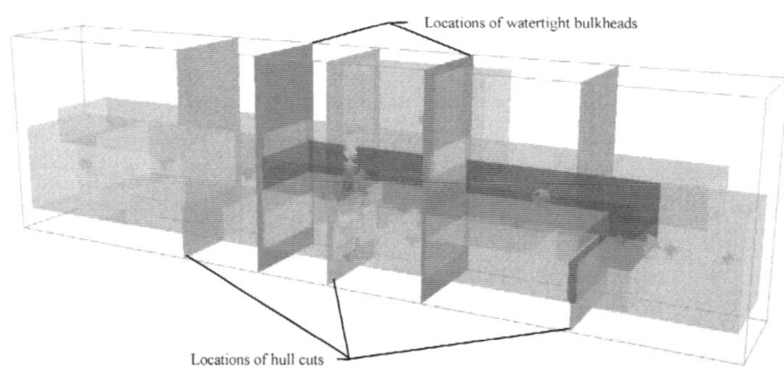

Figure 4. Volume blocks/objects based on the FMFI breakdown (Andrews et al., 1996) with 181 weight items assigned at the centre of each relevant volume block in the initial development of the design (coarse stage).

As the design was developed, the Component Granularity Program (CGP) was available to model features, such as escape towers, torpedo tubes, and other significant components, as necessary (see Figure 5). This level of design granularity shown has traditionally been considered sufficient for typical submarine concept level design. However, since this research emphasised DS3 synthesis for Sustainability, the design was developed beyond this level of granularity, which was facilitated through using the Network Block Approach. This was done by first developing the overall DS3 SUBFLOW network with the four maintenance sections.

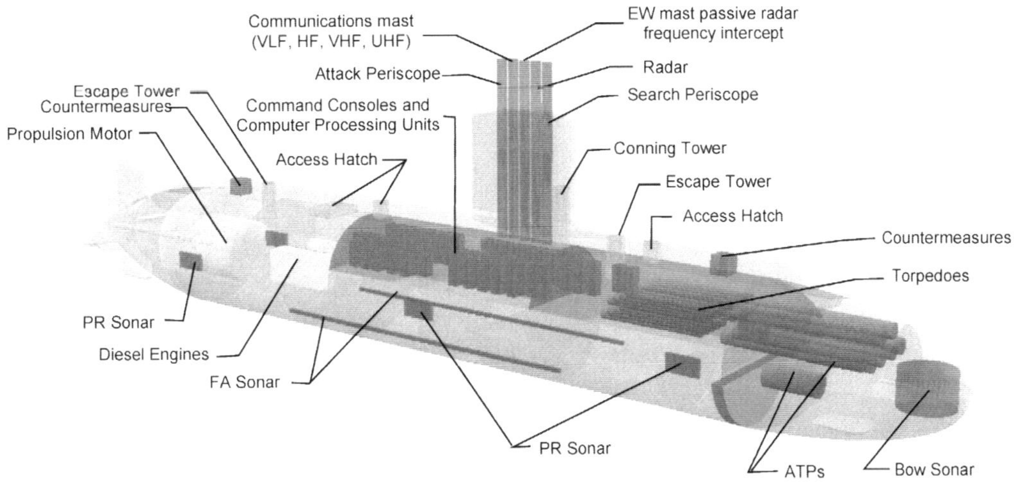

Figure 5. Design progress at the coarse stage showing volume blocks are be included in the hull with major components.

The development of each DS3 network required an understanding of the individual DS3 technologies and how the physics drove a different formulation for the network models. The architecture of each DS3 was also dependent on detailed system types, technology changes and styles of redundancy at the component or connection level (micro-level). Furthermore, when developing the network sketch, spatial aspects, such as zoning, port-starboard, and forward-aft locations needed to be considered. Although these may have needed to be explicitly manipulated in the logical layout of the DS3 network, the spatial order (four maintenance sections) was retained, along with each node's identification.

Using the network sketch the number of DS3 components and connections was identifiable. Thus, relevant equipment data needed to be researched and defined in the Component Granularity Program (CGP). Next, the connections between the DS3 components were defined using the System Preamble Program (SPP) and the System Connection Program (SCP). The DS3 network was then visualised in MATLAB (at a logical representation) and was checked as to whether the network sketch had been accurately modelled.

As the development of the DS3 network was not intended to 'bottom out the preferred design', i.e., it required openness to revision, simplifying assumptions were made for each DS3 technology. This kept the network simple for the ESSD application while avoiding the loss of a sufficient level DS3 complexity definition. Thus, detailed components, which were not explicitly modelled in this network, were accommodated using design margins. This input could be manually adjusted using the System Connection Program (SCP) and the Component Granularity Program (CGP) (see Table 2).

In the case study, different DS3 technologies, ranging from cables to piping (for transporting either gas or liquid) were developed. These were the fuel 'FO' as the source of energy being transferred to various DS3 demands: information data 'DT'; power/electrical 'EL'; mechanical 'ME'; heat removal 'HV'AC (IN, HE, EX); hydraulics 'HY'; trim and ballast 'TB'; and high-pressure, and low-pressure air systems ('HP' & 'LP'). The development of each DS3 is outlined in detail in Mukti (2022).

Figure 6 (top) indicates the overall DS3 network developed with four main maintenance sections, while Figure 6 (bottom) shows the physical architecture of DS3 and the location of four speculative circumferential hull cuts. There were 231 nodes and 468 arcs (see Newman (2010) for network terminologies).

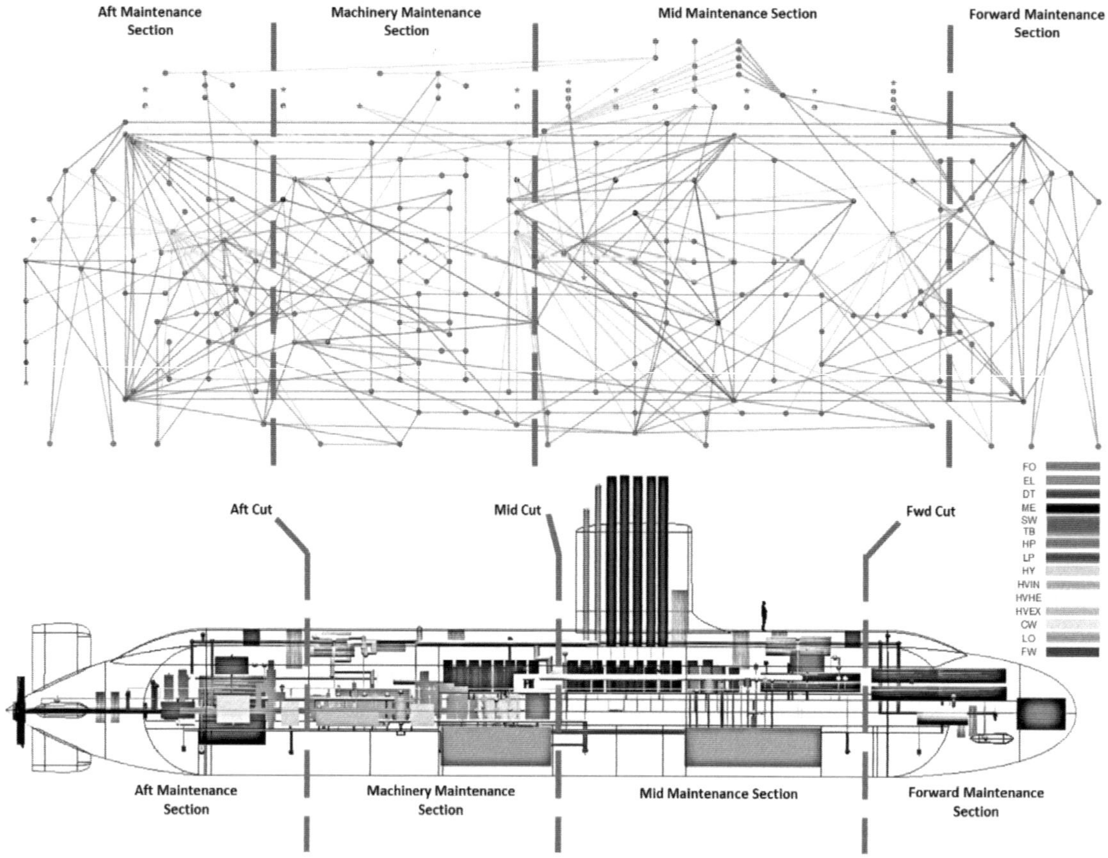

Figure 6. The resultant SSK concept level balanced design solution addressing Sustainability with four maintenance sections shows the overall DS3 network with 231 nodes and 468 arcs. This can be transformed rapidly to physical architecture representation using the programs within the Network Block Approach – see Table 4 for the case study's submarine resultant characteristics

Table 4. Initial submarine resultant characteristics

Variable	Value
Surfaced displacement	2250 te
Submerged displacement	2496 te
Power generator	2 x 1.4 MW diesel engines
Lead acid batteries	480 cells
Submerged BG	0.6 m
Length	65 m
Pressure hull diameter	7.6 m

As shown in Figure 6, the automatic routing algorithm in Paramarine-SURFCON ignores or violates other BB objects. Such a routing clash can be avoided by adding more inputs to Paramarine-SURFCON. However, it can be time and labour intensive. As a resolution, the current approach adopted highway routing framework, commensurate with ESSD (see Mukti (2022)). The design process of the case study is summarised in Figure 7, which involved different types of initial submarine design balance: weight balance; space balance; longitudinal balance; stability balance; and energy balance.

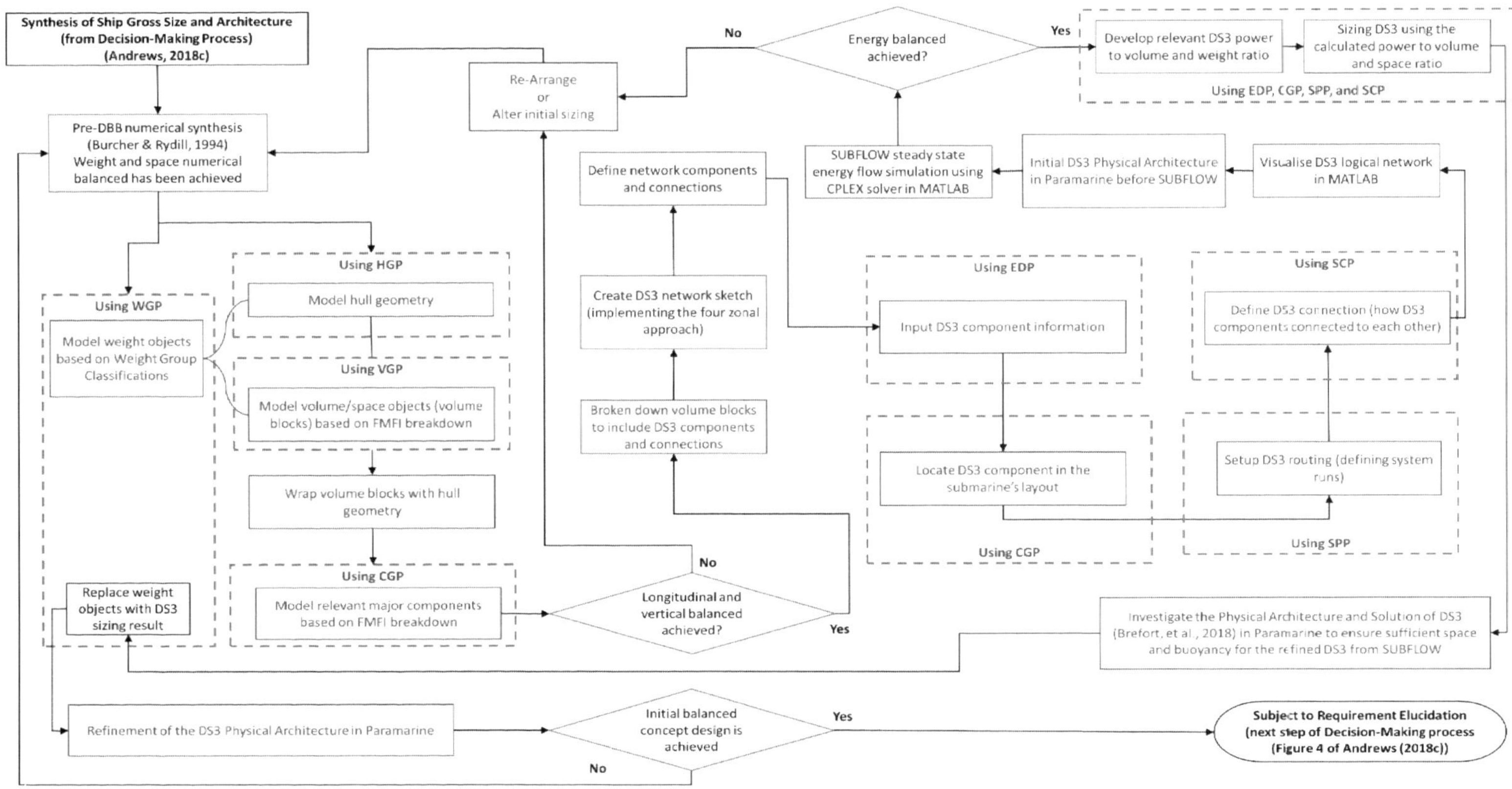

Figure 7. Detailed proposed Network Block Approach (NBA) procedure for Sustainability, which consists of balance check gates (red), the Physical Loop method (purple), input data centre (green), and Logical Loop method (blue) (see Mukti (2022) for detailed explanation of the NBA procedure).

4.2 ADDRESSING THE IMPACT ON THE DESIGN

The impact of implementing Sustainability considerations on the design was quantified in terms of weight and space, which is discussed in the next two subsections in turn.

4.2 (a) Weight Impact

The DS3 weight inputs for the case study calculated using SUBFLOW in the Network Block Approach are compared to the DS3 design that was estimated using the displacement-based algorithm approach. The latter was based on fictitious yet not unrealistic submarine design data (UCL-NAME, 2014). In summary, SUBFLOW could be used to give about 50 weight and space inputs to the design. However, there was no available data/algorithm to compare individual volume results and thus only weight was compared in the study.

There were no significant differences between the two approaches for Weight Group 2 classification system using the UCL-NAME data except for the main batteries and the propulsion motor, which were about 5% different from the UCL parametric estimate. However, after introducing a numerical node to allow the heat to escape into the environment in the SUBFLOW formulation, the difference was reduced to 0.7%. The overall weight of the electrical system for Weight Group 3 shows a relatively small deviation from the scaling approach estimate in that it only differed by some 4.8%. For Weight Group 4, there was also no significant difference, from that for Weight Group 2 (~3 tonnes). However, in percentage terms, the difference was likely to be higher, at 9.3 %.

Essentially, the majority of DS3 components lie in Weight Group 5 where it was found that the result from SUBFLOW sizing was lower than the displacement-based scaling approach by 20.7% or 15 tonnes (see Table 5). Furthermore, although the size of the fuel tanks could be determined by simple numerical calculation, the fuel nodes could be used to refine the size of the fuel tank blocks/objects. The derivation of the power to volume ratio of the fuel nodes required more input, such as the specific fuel consumption (SFC) and fuel weight density and thus the difference in Weight Group 9 was about 7.1%. Overall, the difference in the weight between the UCL parametric approach and the network sizing based SSK was 18 tonnes or less than 1% of overall submerged displacement.

Table 5. Comparison of the NBA/SUBFLOW analysis with parametric approach for case study DS3 weights

WG5	Ship Services	NBA/ SUBFLOW (te)	Calculated weight (te) using scaling/parametric approach - general guideline based on the UCL NAME (UCL-NAME, 2014)		
w51	Snort induction system	1.1	4.2	3-6 te	Dependent on mast height
w52	Ventilation trunking	1.6	3.5		Scaled from SSK (UCL Submarine Databook 2014) Weight Breakdown Profile
w52	Ventilation fans control	7.9	1.2		0.2/0.4/0.5 (6 units 0.2 te)
w52	Vent valves filters		1.3		Scaled from SSK (UCL Submarine Databook 2014) Weight Breakdown Profile
w52	Vent heaters		0.1		Scaled from SSK (UCL Submarine Databook 2014) Weight Breakdown Profile
w53	CW plants	2.1	6.0		2 unit of 56 kW CW plants
w53	CW system	11	7.5	$\approx 0.003\Delta_{sub}$	Dependent on submarine size
w53	SW cooling system	2.8	12.2	$\approx 0.004\Delta_{sub} - 0.005\Delta_{sub}$	Bigger ratio for smaller submarines
w53	Trim bilge ballast system	8.5	8.8	8-13 te	Dependent on trim tank size and position
w53	Trim pp starters	2.5	0.5	0.5 te	Assumed fixed value
w53	HP bilge pp starter		1.5	1.5-2.5 te	Dependent on MBT size
w53	LP bilge pp starter		0.2	0.2 te	Assumed fixed value
w53	HP ballast pp starter		1.5	1.5-2.5 te	Dependent on MBT size
w54	FW cooling system	2.2	2.9	2.5-7 te	Dependent on size of submarine
w55	HP air compressors	5.4	5.4		2 units (320m^3/hr)
w55	HP air system	0.6	3.9		Scaled from SSK (UCL Submarine Databook 2014) Weight Breakdown Profile
w55	Direct blow system		2.1		Emergency blow system

WG5	Ship Services	NBA/ SUBFLOW (te)	Calculated weight (te) using scaling/parametric approach - general guideline based on the UCL NAME (UCL-NAME, 2014)		
w55	Aux vent blow system		3.6		Scaled from SSK (UCL Submarine Databook 2014) Weight Breakdown Profile
w55	LP blower	2	2.0		Assumed one blower
w55	LP blow system	3.4	4.0		Scaled from SSK (UCL Submarine Databook 2014) Weight Breakdown Profile
w57	Main hydraulic system		4.5	4-7 te	Dependent on main plant
w57	External hydraulic system	4.7	2.3	2-6 te	Dependent on external plant
w57	Steering hydroplane control system		1.4	1.5-3 te	Dependent on control plant
w57	Steering diving hydraulic plant		2.5	3-4 te	SDB 121 120l/mins dependent on control surfaces size
w57	Main hydraulic plant	10	1.5		SDB 121 60l/mins dependent on required capacity
w57	External hydraulic plant		1.5		SDB 121 60l/mins dependent on required capacity

4.2 (b) Space Impact

As shown in Figure 8, the aft maintenance section contains most of the DS3 components and connections. The propulsion motor, hydraulic plants and HP air compressors were fitted in this section. This architectural decision led to a 1% increase (equivalent to 4 m^3) in the aft maintenance section from the initial packing density estimate. Figure 8 also shows the locations of connection cuts that could be initially identified, which include LP air system, HVAC system, HP air system, hydraulic system, LO cooling system, FW cooling system, information data system, fuel oil system, trim and ballast systems, and electrical/power system. This architectural model further revealed that the propulsion motor could be removed after the HP air compressor and propulsion switchboard removals.

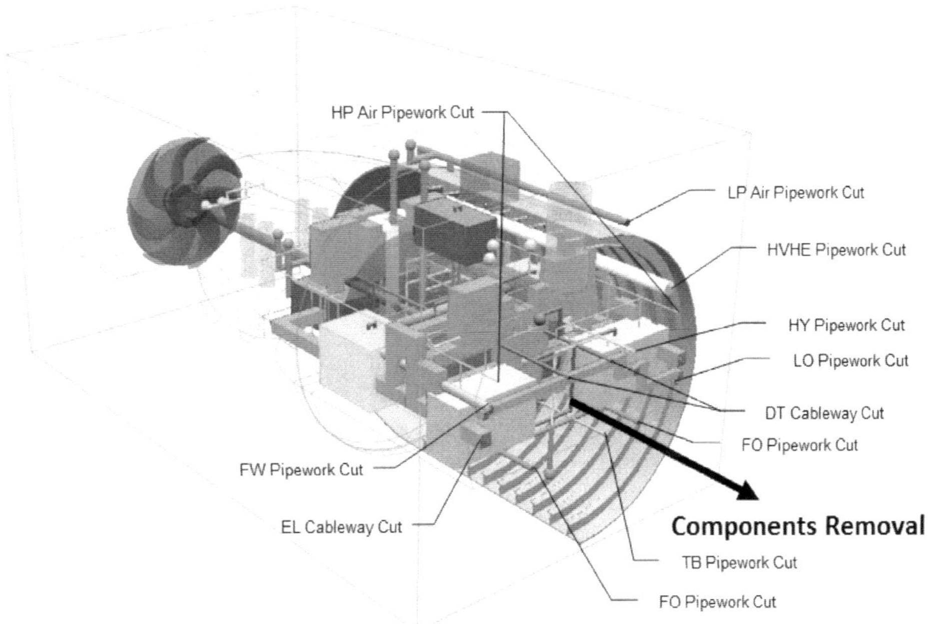

Figure 8. Aft Maintenance Section (with starboard side pressure hull hidden)

The machinery maintenance section (see Figure 9) needed to be increased by 4% (equivalent to 5 m^3) from the initial UCL parametric estimate. In this section, the major DS3 components, such as diesel engines and the aft battery group can be removed for maintenance. The majority of DS3 connections are also affected by the hull cut. These are LP air system, HVAC system, HP air system, electrical system, LO system, hydraulic system, FO system, FW system, information data systems, and trim and ballast systems.

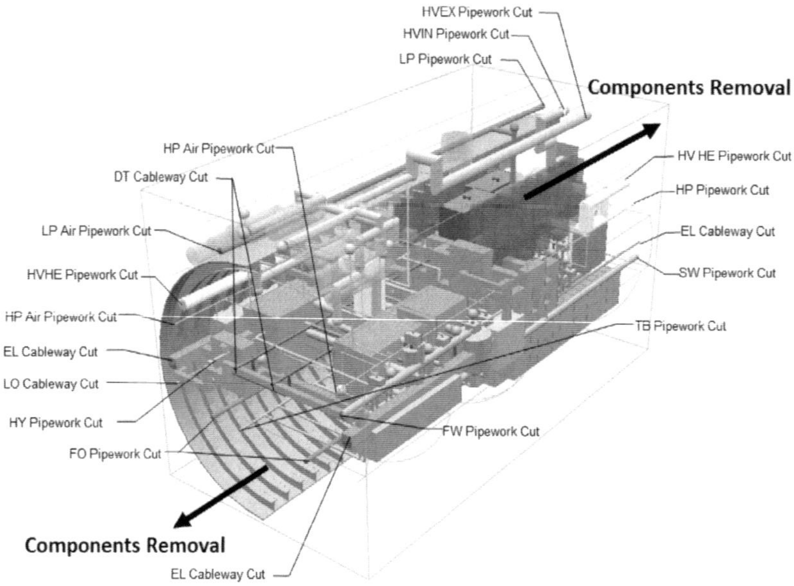

Figure 9. Machinery Maintenance Section (with starboard side pressure hull hidden)

The mid maintenance section was increased by 8% (equivalent to 19 m^3) from the initial packing density estimate to accommodate electrical systems, information data system, trim and ballast systems components that are deliberately separated (forward and aft) due to the style decision. The mid maintenance section allowed the aft battery group to be removed as well as CPUs and command consoles. Figure 10 shows there are HVAC systems, HP air system, LP air system, hydraulic system, electrical system, information data system, trim and ballast systems in the mid maintenance section.

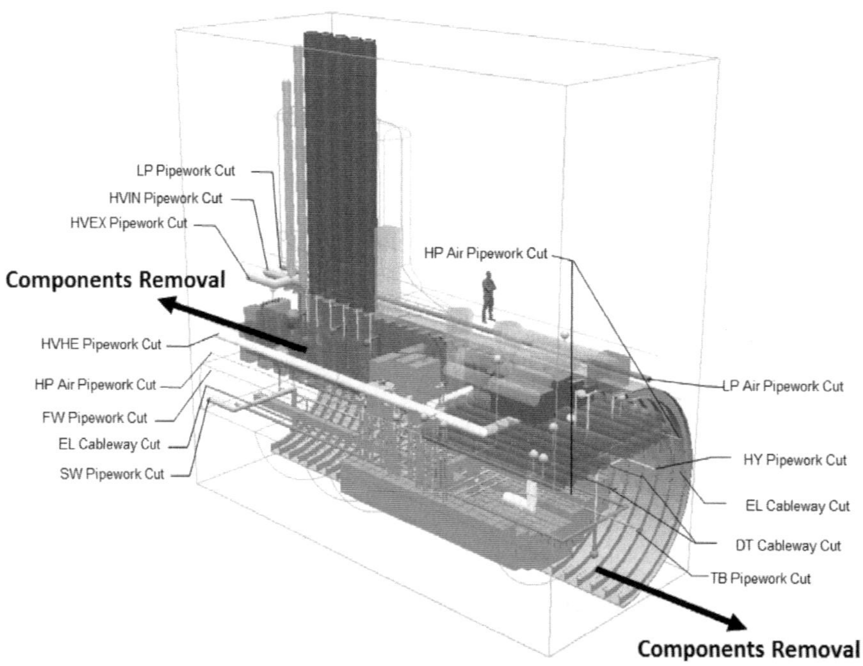

Figure 10. Mid Maintenance Section (with starboard side pressure hull hidden)

The forward trim tank in the forward maintenance section was increased by 10% (equivalent to 20 m^3) from the original estimate using the T&C algorithm (UCL-NAME, 2014) to ensure a correctly sized trim and compensation system for all potential combinations of the variable load and submarine geometry. This section allowed structural maintenance surveying for the forward dome bulkhead with large penetrations for the six torpedo tubes and two ATPs are located. As shown in Figure 11, the DS3 connections that are affected by the hull cut are LP air system, HP air system, electrical cableway, hydraulic system, information data systems, trim and ballast systems.

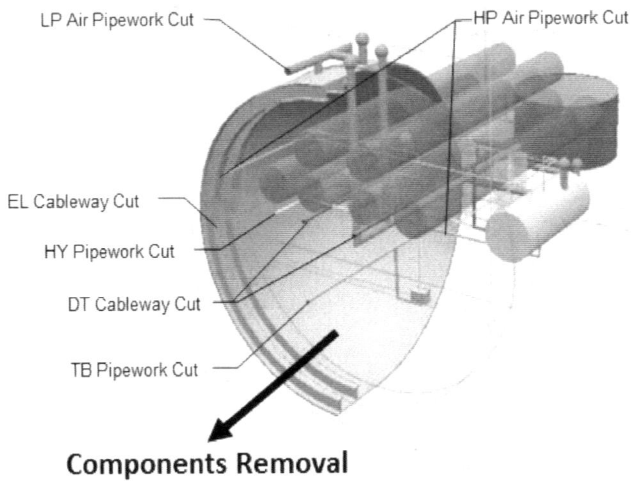

Figure 11. Forward Maintenance Section (with starboard side pressure hull hidden)

Although many detailed DS3 components were not modelled using the Network Block Approach, this study estimated at least 1.5 km of cabling, 280 m of trunking, and 2.8 km of piping are required to be maintained throughout the submarine's life. Furthermore, as multiple hull cuts will introduce structural concerns, this will require further analysis in downstream design. This paper does not encompass the downstream issues of contract design or detailed production design. However, the description of the zonal design of all DS3 facilitated by the Network Block Approach is summarised in Figure 12.

As shown in Figure 12, the lubricant oil cooling system are only situated in the two aft submarine sections. Similarly, the exhaust, chilled water, and saltwater cooling systems are in the middle part of the vessel, i.e., machinery and mid maintenance sections. As expected, the 'mechanical' propulsion system can only be found in the aft part of the submarine. Overall, Figure 12 shows the machinery maintenance section accommodates almost all DS3 except mechanical 'ME' systems. Conversely, the forward maintenance section has the least DS3. It can also be seen that the fuel oil and the freshwater cooling systems are situated in all sections but the forward maintenance section. This was an attempt to minimise DS3 connections transit across the forward hull cut location.

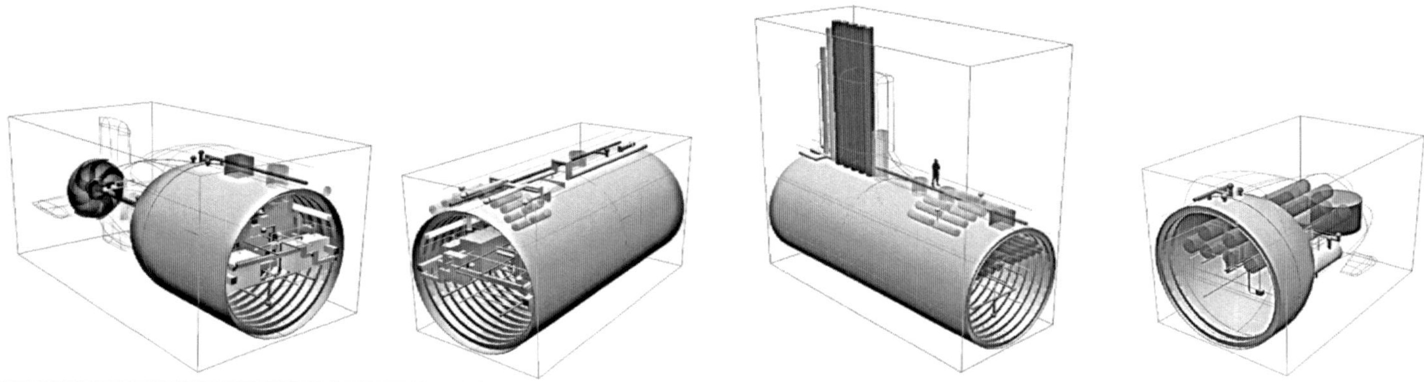

DS3		Aft Maintenance Section	Machinery Maintenance Section	Mid Maintenance Section	Forward Maintenance Section
FO		Yes	Yes	Yes	
EL		Yes	Yes	Yes	Yes
DT		Yes	Yes	Yes	Yes
ME		Yes			
SW			Yes	Yes	
TB		Yes	Yes	Yes	Yes
HP		Yes	Yes	Yes	Yes
LP		Yes	Yes	Yes	Yes
HY		Yes	Yes	Yes	Yes
HVIN			Yes	Yes	
HVHE		Yes	Yes	Yes	
HVEX			Yes	Yes	
CW			Yes	Yes	
LO		Yes	Yes		
FW		Yes	Yes	Yes	

Figure 12. Summary of the zonal approach to the initial DS3 design for the SSK Case Study

5. CONCLUSION

This paper has demonstrated the potential of the proposed Network Block Approach in addressing the architectural aspect of submarine DS3 and tackling supportability issues in ESSD. Submarine DS3 arrangement and physical routing are reasonably complex and thus any 2.5D design tool (see Figure 13) will be unlikely to be able to capture overlapping DS3 routing due to different highways being on the same deck. Furthermore, the physical interactions of DS3 components are going to be scattered in a 3D space within a compartment and, without a 3D representation, consideration of accessibility to those DS3 in the confines of a crowded compartment would be hard to address (see the space impact example in Subsection 4.2 (b) and Figure 13).

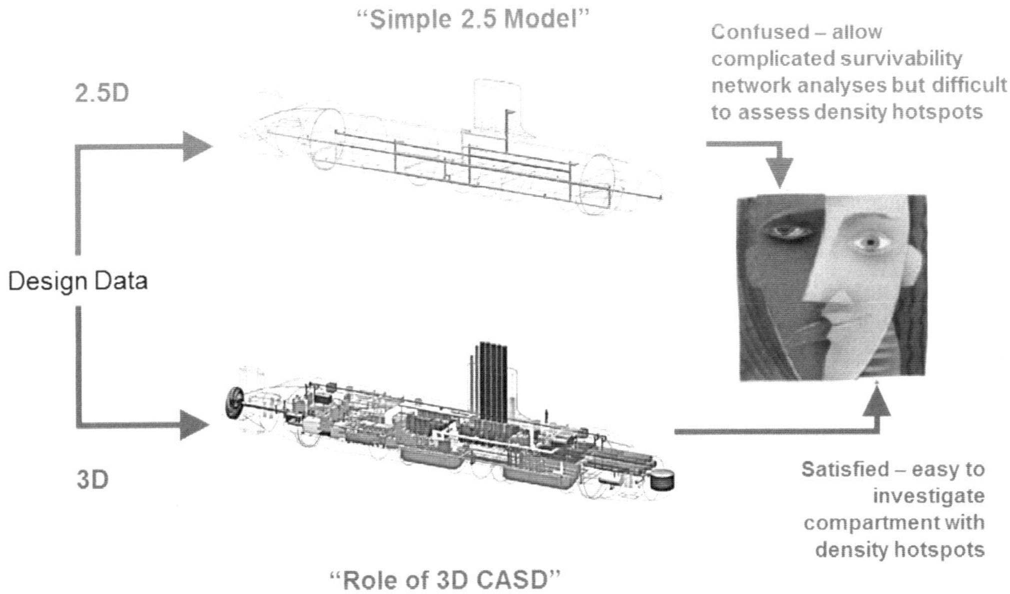

Figure 13. The role of 3D Computer Aided Ship Design in ESSD

The identification of such 3D related issues early in the design process would help ensure the selected DS3 design could be integrated into the submarine design beyond compartment or packing densities assumptions (such as (Purton, 2016)) that need to be based on previous designs. Furthermore, given the advances in computer graphics, not utilising such technology to create a 3D based synthesis was seen to be not taking advantage of CAD developments. The issue is seen rather be how to make the 3D based synthesis execution process as simple as possible so that the designer can readily work on the 3D architecture of the vessel and focus on important architecturally driven decision making in ESSD. When the design approach is 3D architecturally driven, as adopted in the Network Block Approach, the wider style issues like supportability can be investigated early.

Finally, when developing the network-based approach, it was found that a full network analysis can also be time intensive and as demanding as CFD or FEA analyses. Recognising the need to reduce the designer workload in ESSD and be able to readily explore different configurations, SUBFLOW has been devised. It can than be used to employ network analysis to a relatively simple level to enable DS3 synthesis and investigate density hotspots, typical of the 3D nature of the submarine design. Further discussion on the possible investigation using SUBFLOW in ESSD will be addressed in future papers.

6. ACKNOWLEDGEMENT

This work was supported by the funding from the Indonesia Presidential Scholarship and the UCL Joint Scholarship awarded to the first author.

7. REFERENCES

1. ANDREWS, D.J. et al., (1996). SUBCON - A New Approach to Submarine Concept Design, in: RINA Warships Proc. Symp. Naval Submarines 5: The Total Weapons System. Royal Institution of Naval Architect, London, UK. June 1996.

2. ANDREWS, D.J., (2011). Marine Requirements Elucidation and the Nature of Preliminary Ship Design. RINA IJME, International Journal of Maritime Engineering Vol. 153. 2011.

3. ANDREWS, D.J., (2017a). Submarine Design Is Not Ship Design, in: RINA Warship Conference: Naval Submarines & UUV's. Royal Institution of Naval Architect, Bath, UK. June 2017.

4. ANDREWS, D.J., (2017b). The Key Ship Design Decision - Choosing the Style of a New Design, in: 16th Conference on Computer and IT Applications in the Maritime Industries. Cardiff, UK. May 2017.

5. ANDREWS, D.J., (2018). The Sophistication of Early Stage Design for Complex Vessels. RINA IJME Spec. Ed., International Journal of Marine Engineering Vol. 472 Part A. 2018.

6. ANDREWS, D.J., (2021). Who Says There Are No Real Choices in Submarine Design? Presented at the WARSHIP 2021: Future Technologies in Naval Submarines, Royal Institution of Naval Architects, Virtual. June 2021.

7. ANDREWS, D.J., PAWLING, R.J., (2003). SURFCON A 21st Century Ship Design Tool, in: The 8th International Marine Design Conference. IMDC, Athens, Greece. May 2003.

8. BREFORT, D. et al., (2018). An architectural framework for distributed naval ship systems. Ocean Eng. Vol. 147. 2018.

9. BROWN, A.J., (2020). Design of Marine Engineering Systems in Ship Concept Design. Chapter 1 of the Marine Engineering Series 2020. SNAME. Alexandria, VA. 2020.

10. BROWN, D.K., ANDREWS, D.J., (1980). The Design of Cheap Warships, in: Proc. of International Naval Technology Expo 80. Presented at the Proc. of International Naval Technology Expo 80, Rotterdam, Netherlands. 1980.

11. COLES, J., (2014). Study Into the Business of Sustaining Australia's Strategic Collins Class Submarine Capability. Ministerial and Executive Coordination and Coommunication Division Defence. Canberra, Australia. 2014.

12. HELLYER, D.M., (2018). Thinking through submarine transition. Australian Strategic Policy Institute (ASPI), Defence & Strategy Program. Barton, Australia. 2018.

13. HÜTTERMANN, M., (2012). DevOps for developers, The expert's voice in web development. Apress. New York, NY. 2012.

14. IBM, C., (2014). IBM Knowledge Center. CPLEX® Optim. https://www.ibm.com/support/knowledgecenter/SSSA5P_12.5.0/ilog.odms.cplex.help/CPLEX/MATLAB/topics/cplex_matlab_overview.html (accessed 6.29.19). 2014.

15. MATLAB, (2019). MATLAB - MathWorks. https://uk.mathworks.com/products/matlab.html (accessed 7.8.19). 2019.

16. Microsoft, (2021). Buy Microsoft Excel Spreadsheet Software or Try Excel, Free. https://www.microsoft.com/en-gb/microsoft-365/excel (accessed 6.27.21). 2021.

17. MORRIS, D. & SPINNEY, D., (2017). Designing for Support to meet Challenging Submarine Availability Targets, in: Undersea Defence Technology. Undersea Defence Technology, Bremen, Germany. June 2017.

18. MUKTI, M.H., (2022). A Network-Based Design Synthesis of Distributed Ship Services Systems for a Non Nuclear Powered Submarine in Early Stage Design (Ph.D. thesis). University College London. 2022.

19. MUKTI, M.H., PAWLING, R.J., ANDREWS, D.J., (2021). Distributed Ship Service Systems Architecture in The Early Stages of Designing Physically Large and Complex Vessels: The Submarine Case. IJME Vol. 163. 2021.

20. MUKTI, M.H., PAWLING, R.J., ANDREWS, D.J., (2022a). Development of an Early-Stage Design Tool for Rapid of Distributed Ship Service Systems Modelling in Paramarine – a Submarine Case Study. To be presented at the 21th Conference on Computer Applications and Information Technology in the Maritime Industries (COMPIT), Pontignano, Italy. June 2022.

21. MUKTI, M.H., PAWLING, R.J., ANDREWS, D.J., (2022b). The Network Block Approach Applied to the Initial Design of Submarine Distributed Ship Service Systems, in: The 14th Annual International Marine Design Conference. To be presented at the The 14th Annual International Marine Design Conference, IMDC, Vancouver, Canada. June 2022.

22. NEWMAN, M., (2010). Networks: An Introduction. Oxford University Press. 2010.

23. PAWLING, R.J., ANDREWS, D.J., (2011). A Submarine Concept Design - The Submarine as an UXV Motherhip, in: Proc. Warship 2011: Naval Submarines and UUVs. Bath, UK. June 2011.

24. PURTON, I.M., (2016). Concept Exploration for a Novel Submarine Concept Using Innovative Computer-Based Research Approaches and Tools (Ph.D. thesis). University College London. 2016.

25. QINETIQ, (2019). Paramarine. https://paramarine.qinetiq.com/products/paramarine/index.aspx (accessed 7.8.19). 2019.

26. RIGTERINK, D.T., (2014). Methods for Analyzing Early Stage Naval Distributed Systems Designs, Employing Simplex, Multislice, and Multiplex Networks (Ph.D. thesis). University of Michigan. 2014.

27. TRAPP, A., (2015). Shipboard Integrated Engineering Plant Survivable Network Optimization (Ph.D. thesis). Massachusetts Institute of Technology. 2015.

28. UCL-NAME, (2014). Submarine Data Book. UCL Department of Mechanical Engineering, Naval Architecture and Marine Engineering. 2014.

29. WROBEL, P.G., (1984). Design of the Type 2400 Patrol Class Submarine. Trans RINA Vol 1. 1984.

7. AUTHORS BIOGRAPHY

Dr Muhammad Hary Mukti is a researcher at UCL investigating computer-based tools to synthesise distributed systems in the early stages of complex products. Prior to this he obtained an MEng in Ocean Engineering (in the subsea project and underwater vehicle design) from Texas A&M University and accomplished a 2N Professional Summer Course at Massachusetts Institute of Technology. His professional background includes a key person responsible for the laydown area in the detailed engineering phase of the INPEX-Ichthys FPSO project with Daewoo Shipbuilding and Marine Engineering (DSME) and the founder of the Autonomous Marine Vehicle Team Universitas Indonesia that has been actively participating in various competitions held by the Association for Unmanned Vehicle Systems International (AUVSI) and sponsored by the US Navy Office of Naval Research (ONR). In 2022 he was awarded the W H C Nicholas Prize for the best paper published in the RINA Transactions for describing his research in the field of engineering distributed systems design.

Dr Rachel Pawling is a Lecturer in Ship Design at UCL, teaching the subject of ship design to undergraduate and postgraduate students. She obtained her PhD in computer aided ship design at UCL in 2007 and has subsequently continued her research in projects funded by the EC, UK and US governments.

Prof David Andrews, FREng, PhD, FRINA was given a new Chair in Engineering Design at UCL in 2000, following his early retirement as a Director the UK MoD. His early career in the RCNC included postings in in-service and new construction submarine design sections. He was subsequently Head of Preliminary Design responsible for the initial studies, including on the Future Attack Submarine. In 1993-98 he was appointed MoD Professor of Naval Architecture at UCL where he directed the MSc in Naval Architecture and the post-MSc Submarine Design Course. In 2001 he set up a new Design Research Centre in the Department of Mechanical Engineering, which is focused on CASD and the design of complex systems. He is a Fellow of RINA and was a Vice President and was awarded the Froude Medal in 2020. In 2000 he was elected a Fellow of the Royal Academy of Engineering and is currently International Chair of the International Marine Design Conferences.

PRACTICAL SHIP DESIGN CONSIDERATIONS FOR MISSION MODULARITY

A C Kimber and **L Griffiths,** BMT Defence and Security UK

SUMMARY

In this paper, the authors seek to review the direction mission modularity has been taking us and look forward to how modularity will influence future warship design, and some of the challenges that the naval architect will need to address.

1. INTRODUCTION

Modularity has been a practical proposition in naval applications for several decades, mirroring the move in the transportation industry towards standardized load containers typified by the Twenty-foot Equivalent Unit (TEU). Reconfiguring the capabilities of a warship using functional modules was outlined by Spero in 1971 [1]. The Danish Navy introduced the STANFLEX concept in the 1980s with their Flyvefisken class, offering multirole ships, reconfigurable by exchanging modules and others have followed.

The UK's National Shipbuilding Strategy update [2] refers to modularity regarding future RN technologies, stating "Many of these autonomous capabilities and other complex systems will be delivered in a modular manner, which offers the potential to simplify the host platform whilst retaining the flexibility to optimise it for a range of specific tasks. It also provides a route to delivering the adaptability that will be essential for all future Royal Navy ships to enable them to outpace evolving threats and capitalise on emerging technology."

As navies seek more flexibility, is the highly modular warship a viable or even attractive idea, and what adaptations in design will naval architects need to incorporate? Is there a step change in our approach to design required?

2. WHY ADOPT MODULARITY?

Firstly, modularity is a solution, as phrased by Andrews [3] "It could be said that Modularity is a design style solution to the issue of Adaptability" and the RAND report 'Designing Adaptable Ships' [4] further elaborates, suggesting Modularity is a subset of the broader concept of Flexibility and detailing definitions and approaches to both. Flexibility is clearly desirable for a naval vessel but must be considered against cost effectiveness in the real-world use of a warship, as the Figure 1 from [5] illustrates.

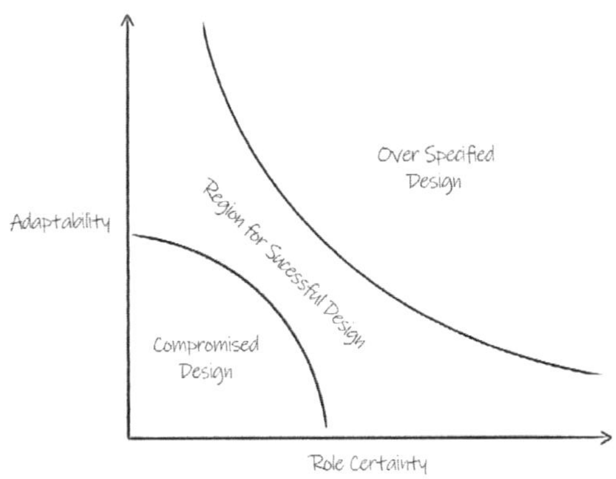

Figure 1: Adaptability vs Role Certainty

Accepted wisdom that survivability is expensive, and modularity is cheap can drive the perspective that all future warships must be modular. However, a deeper consideration suggests this is not a direct conclusion and is a more complex picture. It is important not to fall into the trap of modularity being a solution in search of a problem and a clear understanding from the outset of what capabilities are always required and those that are mission specific is crucial. For example, radar, force protection weaponry, rescue craft etc. will be inherent in every mission requirement and therefore there is little to be gained in making these elements modular. The true question for the designer should therefore be how much compromise is required in the trade between cost and flexibility of capability.

The adoption of modular mission systems can broadly be linked to desire for the following outcomes:

- Allowing an individual ship to perform multiple but discrete missions, creating fleet flexibility and reducing required hull numbers in comparison to mission specific ships. This could deliver a financial saving, in acquisition and operational costs, but critically the missions cannot overlap such that time and planning enables the ship to re-role. It is therefore well suited to intermittent but planned tasking and ideally close to the operating base.
- Allowing capability levels to flex in an uncertain world, adjusting the number of vessels configured to particular missions. This gives a navy flexibility to retain core capabilities and then surge if they are required, offering a saving through fewer hulls although the ability to "surge" capabilities would require modules to be purchased, supported and personnel trained. Again, the tasks need to avoid overlapping as capability will still be limited by hull numbers.
- Technology and equipment updates may be more easily achieved with modules removed for replacement or upgrade on shore. In this case the proposed cost saving is not hull numbers but reduced through life update costs. These need to be considered and traded against the initial investment in applying modularity, so it is a different argument to those above.
- The ability to adjust mission in theatre and rapidly. Here, it is a saving by not equipping a ship for all missions, so the installation cost is saved and potentially some equipment saving may be possible if not all ships are to be equipped at the same time. This is traded against the cost of applying modularity again.
- Modularizing capability simplifies the ship, putting complexity into modules that can be built in factories by specialist suppliers and reducing the ship outfitting costs. Here it is a cost trade within the acquisition and manufacturing process, and it also rests on the ship being able to be simplified which can and probably is over estimated.

From the above, modularity can be about requiring fewer ships, seeking to reduce individual ship acquisition costs, or as a future hedge against changing military needs. Aspects such as mission concurrency, both ship and capability availability and operating pattens all need to be considered. As pointed out by Spero when proposing modularity, the first objective listed was to "determine the potential worth to the Navy of building ships with large scale functional modules".

3. WHAT TYPE OF MODULE?

At its basic level, a module is an independent or self-contained unit, often standardised that can be used to build a more complex whole. In the warship context, it represents an envelope with common interfaces that can be "easily" embarked, removed or exchanged from a ship to change its capabilities. On the basis that the most benefit is gained by being able to put modules on to different ships, this implies a common box (although in the spirit of the above comments, this is a perception of the requirement).

Beyond the basic definition of size and dimensions, other aspects to consider are maximum weight, the need for services and system supply, types and position of service connections, mounting and securing fixtures and access. Whilst not diving into detail in this paper, some of the aspects with the most importance in driving a ship design are considered further below.

The size of a module will obviously have to consider the expected types of equipment to be incorporated and is therefore a somewhat open question. Table 1 illustrates some of the standard container sizes and existing naval modularity units. In fact, a more common approach being adopted is to select the dimension based on widely recognised container sizes and then identify the capabilities that fit; this both bounds the problem and ensures modules can be integrated within existing logistic handling arrangements.

The most widely cited basis for a module is the ubiquitous TEU [6], which has already established standards and can be transported readily. For example, the DNV regulation for Offshore Containers [7] readily provides a basis for the (structural) design, materials and lifting / loading requirements for containers to be used in an offshore environment including handling at sea. Much work has also been done considering how this standard box can be utilised for packaging military capability.

	Length	Width	Height
Twenty-Foot Equivalent Container (TEU)	6.06m	2.44m	2.59m
Forty-Foot Equivalent Container	12.19m	2.44m	2.59m
Tri-CON	1.97m	2.44m	2.44m
STANFLEX	3.5m	3.0m	2.5m
SH Cube	6.06m	2.44m	2.90m

Table 1: Comparison of Standardised Module Units

Within the naval domain, recent years have seen the introduction of more standardisation to enable a coherent approach to modular capabilities [8, 9], particularly in terms of general construction and interfacing of containerised systems resulting in the issue of the Design and Interface Standards for Containerised Mission Modules, ANEP-99 [10]. This standard includes referenced material from the DNV regulation for offshore containers but expands to cover aspects relevant to the naval environment. This provides a framework for standardisation of the containers themselves, although not how they are used within a design. Alongside the size of a module, another key aspect to consider is how the capability interacts beyond the module boundary. Does it require access or need equipment movement through the sides or the top, does it need to be on the open weather deck or within an enclosed environment and are the contents likely to be "wet" or "dry"? These choices fundamentally impact how the modules are arranged into the ship design and are not complimentary.

The integration of the module needs to be considered and is driven by the selected module standard. A TEU is secured at its lower corners using standard connectors, although other suitable naval connectors including for shock have been devised. This approach offers simple and quick embarkation and a great flexibility in how the TEUs may be arranged. However, the container is in this case not structurally integrated with the ship requiring careful attention to the load paths through the securing feet depending upon the nature of the capability hosted. It is notable that the STANFLEX modules chose a different approach, as these are intended to slot into the deck of a ship and are therefore secured at their top edge, using bolts, such that they hang from the deck. This approach is more in keeping with the STANFLEX concept which included gun and missile modules. However, this style of module introduces the need for more careful attention to the tolerances of construction, to ensure that the modules are interchangeable (Spero again recognised early the importance of tolerances).

Table 2 provides an illustration (though not a detailed analysis) of typical capabilities which may be considered for a modularised approach and some of the characteristics that these capabilities may drive. These characteristics will have an impact on the nature of the module (e.g. access or openings required, level of boundary protection), where the module is placed (e.g. wet or dry environments, enclosed personnel access, maximised arcs for firing or sensing) and what supporting services are required (e.g. power, data). Clearly not all modules are interchangeable and for a given design the nature, if not the specific design, of the modules proposed will have to be defined. It may even be the case that different locations or even types of module may be proposed to match different module characteristics.

Potential Characteristic	Capability: Mission Boat	Aviation	HADR	Accomodation	Weapon Launcher	C2 Space & SCIF	Offices, Workshops, Labs	Towed Sensor	Self Defence System	Medical Facilities	Magazine	Sensor	UUV	Cranes	Minelaying	Hydrographics Equipment	Bulk Stores
Wet Environment	■							■					■		■	■	
Dry Environment		■	■	■		■	■			■	■	■		■			■
Open Deck Environement	■				■				■			■		■			
Hazardous	■	■			■				■		■		■		■		
Personnel Safe Environment				■		■	■			■							
Walk in Access			■	■		■	■			■							■
All Round Access	■	■			■				■				■	■	■		
Power Required	■			■	■	■	■	■	■	■		■	■	■		■	
Data Link Required					■	■	■	■	■			■	■			■	
A/C Controlled Atmosphere				■		■	■			■	■	■					
EM Environment Implication						■					■	■					
Dynamic Loads	■				■			■	■				■	■		■	
Requires Multiple Modules				■					■	■					■	■	

Table 2: Module Characteristics versus capability

4. WHAT IS THE BEST WAY TO ARRANGE MODULES ON THE SHIP?

When considering how to arrange modules into the ship design, we can broadly categorise three approaches as shown in Figure 2. Each of these approaches offers different advantages matching different module characteristics as discussed above:

- Use of a mission bay offers flexibility, including capabilities requiring multiple modules or more flexible use of the space for different module sizes or non-modularised equipment. It provides shelter, protection and ease of access, but is not appropriate for "effector" type modules and complicates aspects of handling modules and deploying equipment.
- Using an open weather deck to stow modules is less complex in terms of handling modules and opens the potential for modules which require open air, such as weapons or sensors, but protection from the elements is reduced making the option less attractive for accommodation or office type applications.
- Distributed modules allow the designer to place individual modules where they are based suited, an ideal solution for sensors and weapons but the space is more constrained and less flexible.

Figure 2: Examples of distributed modules (top); an enclosed mission bay (left) and weather deck modules (right)

A mix and match strategy would offer a good compromise, but the implications on the ship design become exaggerated when trying to incorporate multiple module applications. It is therefore key to understand what types of capabilities would likely be embarked to select the most appropriate method of incorporating them into a ship design. If the capabilities are largely associated with stores, equipment or additional working space then grouping within a mission bay may be preferred, whilst if considering sensors or weapons a distributed modular approach may be best. Unfortunately, this may also make the standardisation of modules between different types of ship less achievable or attractive; is it reasonable or sensible to have a common module for a weapon that fits into a specific slot on a combatant as a container to carry humanitarian stores on an auxiliary vessel? This perhaps suggests that a modular ship may not necessarily mean a ship arranged to accommodate a single type or size of module; rather it may employ several module concepts for different uses or even other forms of adaptability, such as illustrated in Figure 3.

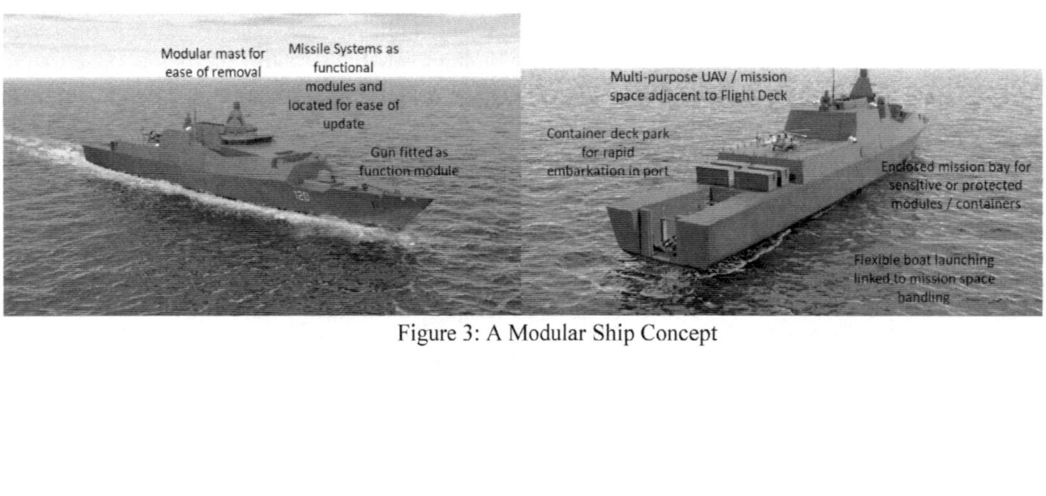

Figure 3: A Modular Ship Concept

5. OFFBOARD SYSTEMS AS MODULAR CAPABILITY?

There is an increasing trend in the use of off board systems, and these are in their nature "modular". Being able to change systems requires the supporting systems including launch and recovery to either be common or themselves modular. For aircraft, the ability to embark different helicopters is already mature thinking; providing a large enough hangar and flight deck strength for flexibility is straight forward. The supporting equipment and stores are less easy as each helicopter will have its own requirements, but the adoption by the RN of standardised bulk stowage modules fitted to bulkhead racking has achieved this aim. Progress in terms of surface craft is not as far advanced, although flexible handling systems have been developed, but will be an additional cost to the base ship.

Figure 4: Flexible launch and recovery and boat handling systems are now being marketed

The impact that a more off-board system-oriented approach will have on the future design is going to be driven by the combination of the number and the size of the chosen vehicles. Unless only small vehicles are required, then the space and weight impacts will quickly be multiplied to dominate the design. With most USV craft in the region of 7 to 12 metres length, and with at least two, more realistic at least three or more craft necessary (for redundancy, availability and mission duration reasons) this is a significant real-estate requirement.

It can quickly be observed that an offboard focused warship would require large internal garage and hangar spaces for the vehicles, noting helicopters do not like the same environment as USVs so probably two distinct spaces. Hence, the concepts for mothership tend to be of not insignificant size; this is not necessarily an issue if they can be in the support ship category and therefore of more commercial build. But if intended to operate in opposed and higher threat environments then the viability of a low-cost mothership is questionable as from the survivability perspective the mothership and its assets are a system and survivability trades must consider both (for example the loss of the Atlantic Conveyer was costly to the operation because of the helicopters lost on board, not the physical ship itself).

As well as vehicles, it is also worth considering launchers for missiles within this discussion. It is possible to modularise specific weapon launchers and therefore different type of weapons could be embarked for different missions, or indeed swapped for non-launcher modules. This has been achieved for some missiles systems, notably the ESSM launcher used as part of the STANFLEX range. However, an alternative path is the multi-role VLS. Systems such as the Mk 41 VLS can launch a range of different missiles for different capabilities, without the need to exchange the launcher and its supporting systems. They are also in themselves considered modular, being self-contained units, although this is more associated with installation ease than capability exchange. This demonstrates how a design might apply different forms of modularity for different capabilities.

6. WHAT ARE THE IMPACTS ON SHIP DESGIN?

The adoption of modules to deliver mission capabilities can have profound impacts on the design of a ship, which quickly escalates with the number of modules considered. Spero [1] outlined the impacts at the earlier stages of the debate on modularisation; maintaining watertight and structural integrity, controlling weight, centre of gravity variations, compatibility with the arrangement configuration, support service interfaces. Even this early Spero et al recognised that "inherent to the concept, however, is the expectation that ships with large modules will be larger and more expensive in initial acquisition than comparable conventional ships". This returns us to the earlier question of what is being traded – capability on the ship at a given time or number of ships?

There will be an impact on stability associated with the combined effects of the potential higher payload weight for modular systems and the impact that any mission module space may have upon the watertight integrity and buoyancy. A mission space that is readily accessible may need to sacrifice watertight integrity and its size and location will vary the impact it has on stability performance. When considering purely commercial or benign roles and hence commercial standards larger and more accessible spaces are achievable (for example platform supply vessels with a large open working deck) but applying naval damage standards is already a challenging exercise for the designer without consideration of mission bays.

This is illustrated by a short study into the impact of different configurations of a ~130m naval vessel. As shown in Figure 5, six configurations were tested with an increasing length of mission space forward of a stern ramp and also options with and without outboard watertight spaces. In each case an increasing amount of buoyancy reserve is lost as the mission space is assumed as non-watertight.

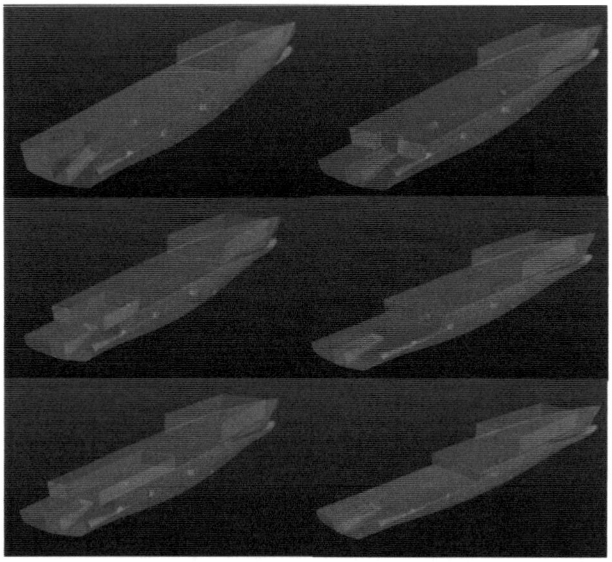

Figure 5: Configurations of buoyant volumes for mission spaces

From top row left to right - a stern ramp, a stern ramp with open decks extending width wise, a stern ramp with a mission bay forward but watertight spaces outboard; similar but with a full width mission bay; an extended mission bay with outboard spaces and extended without outboard spaces.

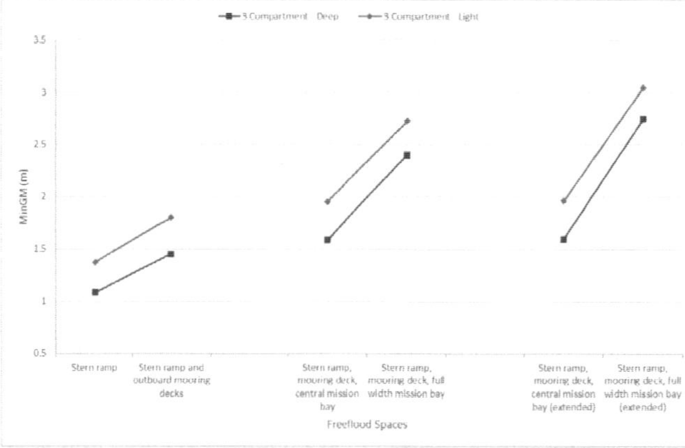

Figure 6: Comparison of Stability for different non-watertight mission spaces configurations, expressed as required GM

The results are illustrated in Figure 6 and are expressed in terms of the required GM to meet naval stability criteria, assessed for comparison against a three-compartment damage standard in deep and light loading conditions. These results show that as expected, the increasing loss of buoyancy reserve drives the stability, as shown by the greater required GM for compliance. What can also be observed is that this is far more pronounced when the mission spaces are full width. This is because the mission spaces are above the intact waterline and as such the lost buoyancy is most important when heeling. Therefore, it is the outboard volume that has the greatest impact on stability performance. Provided the mission space is at a suitable height above the waterline (in this case above the damage control deck) and buoyancy outboard is provided, these large spaces perhaps surprisingly appear possible (although consideration maybe required for free surface effects).

Use of modules will impact ship structures in a number of ways. The addition of a large mission bay would require the following structural impacts to be considered:

- Deck strength and supporting structure (bulkheads) below to address module weights and induced forces, noting these may vary as different modules are embarked and movement may have to be considered).
- Structural integration and continuity where large mission spaces may lead to unsupported deckhead structure and high, possibly double deck height spaces. This may drive larger scantling dimensions and girder depths for decks over or side structure.
- Supporting structure and load cases for any integral handling systems. These may compound the scantling increase further for deckhead structure if overhead.
- Larger access opening in the side shell, longitudinal bulkheads or hatches in decks for movement of modules resulting in loss of continuity of structure.

The design of structure will require the definition of load cases, which may constraint the weight of modules or alternatively require structure to be designed for the worst case and potentially apply margins. Whilst structural issues can be dealt with and designed out or around, the overall implication is likely to be greater structural weight. This in turn is likely to result in greater dimensions to maintain a sensible design balance.

Whilst such design issues may be overcome, it suggests that the arrangement of the ship and its structural style should account for any larger, open modules spaces as part of the initial concept to ensure an efficient design and that weight and space for the required structure is suitably accounted.

Ease of module exchange and access is dependent on the chosen arrangement of module. In the simplest case, modules placed on an open deck are easily embarked or removed by crane from alongside. This of course requires the modules to be resistant to the environment and may impact on aspects such as signatures or vulnerability of the modules. Where mission spaces become enclosed or partially enclosed, then more specific handling systems on board are required. Providing vertical access, for example by an access hatch, remains the simplest method allowing alongside cranage to directly place modules into the mission space. If the space is larger than one module, then internal handling will be required. In its simplest form this maybe skid rails and winching (often used in commercial ships) or a more complex lifting arrangement may be preferred.

Where vertical access is not possible, bearing in mind for naval ships the mission space may conflict with other topside needs (for example a flight deck), then a handling system capable of embarking and moving modules may be specified. This provides a high degree of self-sufficiency, but system complexity can increase rapidly. At the higher end of performance would be a system such as adopted on the Type 26, which can be used to embark TEU's, launch or recover boats and handle both to stowage locations within the mission bay.

It must also be remembered that the modules may require access themselves, space to access at the ends or sides and to remove any equipment if necessary. A large mission bay may also disrupt overall ship access routes depending on its placement.

The modules may also require systems to support, principally electrical supplies but also potential fluids, air and data as well. These systems will need to be designed for flexibility. These for example may require the electrical load chart to account for power to all modules best on a worst-case assumption. This may drive additional load, but issues such as quality of supply and system balance depending on the number and type of modules embarked at a given time. In passing, there may also be implication of the EM environment of the ship.

One key point that becomes apparent from the above discussion is that modularity will drive additional weight into the design and his is proportional to the extent of modularity required due to:

- The weight of module structure and foundations.

- Allowing for maximum module weights rather than specific equipment weights.
- Additional structural weight associated with large span mission spaces.
- The weight of any additional handling solutions.

Together with the need to allow additional space for standardised module envelopes and potential space for access and handling, this clearly implies that modular solutions would drive ship size and displacement and would be greater than an optimised non-modular solution with the same capability. These impacts are demonstrated and quantified through an example extreme modular ship at [10]. Hence modularity offers simplified installation and clearer functional boundaries which may reduce costs, but at increased vessel size which may increase costs. As RAND notes in its study [4] it is unclear if design and acquisition costs of the system, ship and modules, increases or decreases.

7. HOW DOES IT IMPACT SURVIVABILITY?

The modularisation of capability on a warship creates a vessel where survivability, as a function of susceptibility, vulnerability and recoverability is altered by the addition or removal of modules. This introduces a risk of divergence between the ship's survivability as designed, when fitted with modules and as understood by the operator. When designing a ship that will derive combat capability through modules designers will need to understand the 'theatre entry' capability provided by modules. This is similar to how CIWS are not fitted to all Royal Navy vessels at all times but are prioritised for ships on deployed to higher threat areas.

Designers will also need to remain alert to the risk of mission-creep where proposed modules may provide a warfighting capability which is not coherent with the ship's survivability levels. When understood from an early stage and articulated for example within the CONOPS, this will allow the designer to ensure the ship is provided with the appropriate level of survivability for all envisioned modules.

- If this is foreseen at an early stage, the ship can be built with greater survivability than required for its core role to ensure that it does not become a vulnerable asset in high threat environments.
- Regardless, there will be a need to quantify and understand the ship's survivability to ensure that during planning and operations the risk to the ship and its crew is considered.
- If a module is to be used on multiple classes of ship the survivability performance and specification will need to consider high threat and low threat environments.

7.1 SUSCEPTIBILITY

There are several risks to the overall susceptibility of the ship, examples of the risks to ship signatures include:

- Vertical surfaces and microgeometry of externally mounted modules increasing the ship's overall radar cross section.
- Control stations for offboard assets increasing the ship's EM emissions and thus detectability.
- Additional sources of noise and vibration, especially on hard mounted modules increasing underwater radiated noise.
- Inconsistent insulation and additional heat generation leading to an increased IR signature.

Interactions between modules and the ship may also increase the signature beyond a sum of all parts. There are mitigations to the risks identified above that ship designers can incorporate. For instance, external modules can be shielded behind structure or located behind doors to shield their geometry from radar. This will add cost and complexity to the ship. However, if the risks identified are not mitigated then the resulting increased signatures could compromise the ability of the ship's countermeasures to defeat incoming weapons further increasing the overall susceptibility.

7.2 VULNERABILITY

Just as with susceptibility, modules present a risk to vulnerability. However, unlike susceptibility the impact on vulnerability is weighted towards the capability provided by modules. This is for several reasons.

The vulnerability of core functionality such as watertight integrity, damage control or movement should not be affected by the presence of modules on a ship. Instead, the vulnerability of capabilities provided or augmented by modules are those most affected by design choices.

A key challenge presented by modules not carefully integrated within a warship is their damage footprint. Some modules are fully self-contained only requiring power and other ship's services to function. However, the connections to modules represent a single point of failure. Ideally these systems would draw from redundant distribution points or be immediately adjacent to the module but depending on the module's location this may not be practicable. Module slots designed for capabilities with lower vulnerability requirements might not even be considered as requiring redundant supplies.

The protection of modules against shock is another area where the capability is risked. Modules themselves can either be rigidly or shock mounted. If the module slot has sufficient clearance for shock mounts, it would still require a module with appropriate weight distribution and flexible connections for ship services. Alternatively shock mounting can be provided within the module but this limits the available volume for equipment.

7.3 RECOVERABILITY

As with the other aspects of survivability it would seem preferable that modules meet an equivalent standard of recoverability to equivalent fixed installations. Depending on the function of the module this may be necessary or relatively simple such as for modular accommodation or a davit respectively. However, this goal may not always be achievable, therefore designers will need to consider how damage can be fought and contained in and around modules. This includes consideration of:

- Whether modules should integrate into the ship's smoke and fire detection systems.
- Whether fixed fire-fighting systems should be included.
- Access for boundary cooling.
- Conduct of re-entry into modules.
- Use of alternative means such as water-lances.
- Primary and secondary escapes from modules.

A key benefit of modularity is that when recovering capability, it allows damaged modules to be quickly removed and replaced or if the ship has been damaged for functional modules to be removed and installed on other ships to rapidly regenerate the capability.

7.4 SUMMARY

The primary challenge a warship designer faces when considering the impact modules have on the survivability of a ship is the variation between modules. For each module that may be installed a different approach to the control of signatures, management of system dependencies, shock, armour, blast, or firefighting compatibility may be required. Under these circumstances designers may either introduce additional margins, potentially overengineering the design for a capability that will rarely be fitted or ignore the impact of modules entirely. The latter option while simpler introduces a risk that the operator is given an inflated view of the ship's survivability when fitted with modules.

8. HOW DOES IT CHANGE SUPPORTABILITY?

A key feature of adopting modular systems is the ability to support equipment ashore rather than in-situ. This has significant technical cost benefits, including allowing equipment to be returned to the manufacturers for maintenance, where they can be in controlled environments and enable ease of access. Benefits in flexibility and increased availability can be achieved if modules can be rotated to different ships with equipment deployed where needed.

This does create the need for a robust logistics chain to be established; modules must be removed from ships, transported to warehousing and maintenance facilities, stored and maintained in readiness whilst awaiting delivery to a ship. Use of standardised modules sizes such as the TEU aids the logistical chain as they can be transported using existing infrastructure but for deployed vessels there is a need to recognise if these modules can be securely transported over distance, possibly through a number of countries and securely stored overseas if required. There will also be cost involved, particularly if the logistics chain is to be secured "ready to use when needed".

It is well understood that the expected service life of many of the component systems maybe far shorter than the life of the ship so the argument of modularity for future upgrade applies. Such an approach offers a more effective routing to replacing key equipment's, with less intrusive impacts on the ship (for example less hot work) which should reduce time in hand and therefore offer availability advantages. This is of course predicted on the assumption that future equipment would be compatible with the chosen module constraints, and that the physical changes drive the time for refit (and not for example software updates). The diversity and extent of significant upgrades is illustrated by the changes made to the Type 23 frigates, summarised in Figure 7. In this case, it can be deduced that a functional module approach as originally conceived by Spero [1] could be used, more akin to the STANFLEX concept than the containerised TEU as the module may only be exchanged once or on a few occasions and in a shipyard environment.

Figure 7: Type 23 showing selection of upgrades conducted (Picture ©MOD)

Another more specific debate is the length of ship life versus its flexibility. Previously the RN has acquired ships based on a shorter life and more limited upgrade, although this hasn't been borne out in practise. However, it is worth noting at this time there are many innovations expected in the coming decade (net carbon zero, automation, digitisation) and the next generation of ships may either need extensive upgrades as these technologies mature or may have shorter lives as obsolescence is accelerated by disruptive technologies. Therefore, do we need more adaptable ships or shorter lives? But not both?

9. CONCLUSIONS

Is the fully modular ship "as a truck" concept a viable approach? The idea of a warship which is a simple ship, able to embark multiple containers providing a wide range of capability options, appears attractive but is perhaps an oversimplification. It is possible to achieve this provided the module options reflect similar characteristics and uses. Ultimately a highly flexible ship covering a wide range of naval capabilities through modularity may prove challenging and expensive, and delivering greater utility could be more complicated and expensive than initially perceived.

Adopting a high level of modularity for a warship is clearly a significant design driver. Whilst in the short term there is a perceived benefit in flexibility and ability to adopt fast moving technical capability through life, it is harder to optimise as key elements of the design are not fully defined. The designer has to consider all the potential characteristics of modules that may be embarked through the design life and from this distil the driving characteristics. Standards must be considered against each capability option to ensure compliance. It's also likely the ship will be driven to a larger size and care has to be taken when extrapolating from previous non-modular designs.

Clarity on what will be modularised is imperative from the outset with a clear understanding of, when missions will be concurrent and how many modules would a mission require. Hence, the definition of mission payload is a prerequisite to the ship concept which may seem counter to the intent of a modular ship. However, attempting to design a ship with generic module spaces and only limited thought on the size and configuration of mission payloads may result in an ineffective arrangement, equivalent to designing a non-modular ship without early consideration of its weapon systems.

Two different approaches emerge for modularisation. On the one hand a simple standard box which can be secured to a deck and connected to ship systems, typified by the TEU container which can be used to host equipment, people or stores. Or alternatively a more complex functional module which is more integrated within the ship but still is a self-contained capability for easier exchange, such as the STANFLEX module. Both approaches have been developed and offer differing advantages depending on the nature of the capability to be modularised. In fact, they might both have application in the future within the modular ship, representing different layers of modularisation.

By not requiring the installation of complex systems at build and replacing them with standardised interfaces then this implies a simpler construction, particularly when considering that installation may normally occur late in build. But if we consider the ship as a system including relevant modules then overall it is not obvious that the "ship system" is simpler. Many capabilities require not only equipment but impose transversal requirements on the ship, for example speed, shock, signatures, and seakeeping. These transversal requirements apply regardless of the use of modularity and by proposing more flexible capabilities this is likely to increase overall complexity.

These conclusions are driven by the recognition that naval capability is more than the equipment which can be placed in a box. The ship and its modules are a system of systems, which interact and impose requirements on each other. A truly modular ship will require significant design exploration and rationalisation, may utilise different modular approaches to deliver different outcomes and when considered as an overall system is far from a simplified solution.

10. ACKNOWLEDGEMENTS

The Authors wish to thank those colleagues at BMT who assisted with analysis, thoughts and reviews.

11. REFERENCES

1. SPERO Cdr J. R., HICKS Lt Cdr W.F., GREENE Lt Cdr D.L., 'A Philosophy of Naval Ship Design and Construction', Naval Engineers Journal October 1971

2. Refresh to the National Shipbuilding Strategy, 10 March 2022 Page 56, National Shipbuilding Office (CP 605)

3. ANDREWS D, 'The Multi-Role Combatant – Jack of All Trades, Master of None?', RINA Warship 2019

4. SCHANK John F., SAVITZ Scott, MUNSON Ken, PERKINSON Brian, McGEE James, SOLLINGER Jerry M, 'Designing Adaptable Ships', RAND Corporation 2016 ISBN: 978-0-8330-8722-5

5. MOSS Charles, KIMBER Andrew, 'WHEN IS THE ADAPTABLE OR MULTI-ROLE SHIP THE RIGHT ANSWER?', RINA Warship 2019

6. ISO 668:2020 Series 1 Freight Containers

7. DNVGL-ST-E271 DNV Regulation for Offshore Containers

8. ALVAREZ-MARI J, COLE H K, 'Standardisation of Naval Containerised Modules – Enabling Interoperability Across NATO', RINA Warship 2019

9. MANLEY, D., 'The NATO Drive to Mission Modularity', RINA Warship 2018.

10. Design and Interface Standards for Containerised Mission Modules, ANEP-99, NATO Standard, December 2020

11. PAWLING R J, BRADBEER N I C, 'Pushing the Limit: The Role-less Multi-role Warship', RINA Warship 2019

12. AUTHORS BIOGRAPHIES

Andy Kimber holds the position of Chief Naval Architect for Surface Ship Design at BMT. He has over 30 years of experience as a Naval Architect in the defence industry, covering concept studies, design delivery and in-service support. He has held positions in the QEC design team, Tide Class (leading the design) and one of the Type 31 bid teams in the UK and has been involved in projects for the US, New Zealand, Australia, Canada, Norway and Finland amongst others. He has also been involved in projects considering fleet mix balance, options for capability delivery and procurement approaches.

Lewis Griffiths is a Senior Naval Architect specialising in Survivability at BMT. He has 10 years of experience as a Naval Architect in the defence industry. He has worked on multiple design projects including the Tide Class, HNoMS Maud, Type 45 PIP, Type 31 and bidding for FSS, latterly as the Survivability Lead. He has also been involved in a range of research and thought leadership tasks investigating future warship design.

OBSOLESCENCE MANAGEMENT TO ENABLE SUPPORTABILITY FOR THE AUTONOMOUS WARSHIP

T Flint, BMT Defence and Security Ltd, UK

SUMMARY

Rapid obsolescence is often a consequence of rapid technology development. Modern warships are operating in an era of increased technological development, with complex systems having a redevelopment cycle significantly shorter than the lifetime of the ship. The Autonomous Warship concept is one such area and designers must take this rapid development into account to ensure Navies remain at the forefront of technology without early systems experiencing a tsunami of obsolescence. Obsolescence must be considered early in the life cycle of AI and its connected systems to reduce the amount of redundant technology in the future. Advances in AI and mechatronics result in capability improvements, but also increase obsolescence driven waste and supportability issues. This means that effective, proactive obsolescence management is more important than ever to maintain capability, reduce environmental waste and eliminate unnecessary costs. This paper will seek to outline options to increase system adaptability and reduce the looming obsolescence risks of technological advancement.

1. INTRODUCTION

Obsolescence is defined as 'the transition of an item from available to unavailable from the manufacturer in accordance with the original specification' *BS EN IEC 62402:2019*. While items are in this state it is incumbent on system owners to ensure appropriate provision is made to prepare for the future, ideally having planned for this at the outset of the system's life cycle. Often this is managed reactively i.e. waiting for issues to occur before any action is taken, however the pace of development is increasing, which leads to much shorter equipment life cycles, particularly noticeable in emerging technologies. As a result an increasingly proactive approach is needed by both designers and system owners, starting in the early stages of the system life cycle. Key to Navies successfully adopting developing technologies like Artificial Intelligence (AI) will be their ability to constantly adapt and drive change from within. Developing technology and data capabilities to stay at the pioneering edge of this new dimension of automation will be a fundamental building block of a modern and technologically advanced Navy. This capability paired with adaptable equipment will separate those Navies able to sustain this pace of change over the coming decades.
The issue of rapid obsolescence is likely to present itself most keenly to the early adopters of AI. Obsolescence managers will not be able to tackle the huge demands for equipment replacement if reactive management methods are used. Not only will this place ships at a technological disadvantage, but it may also result in reduced availability, noting that a measurable amount of a frigates time is currently spent tending to unplanned repairs[1]. Arguably even more important than this is the impact of wasteful process on the environment. To become more sustainable the Royal Navy developed a Strategic Approach 2021-25 for Sustainable Digital Technology and Services (SDTS). The goals of the SDTS are to reduce consumption of energy and materials, improve digital capabilities to enhance eco-performance, increase supply chain resilience and integrity and increase awareness within the workforce. Managing obsolescence plays an enormous role in supporting these goals; however, the most effective method of managing obsolescence is to reduce its impact through design, starting early in the life cycle and at the lowest practicable component level.

2. THE FUTURE IS AI

AI will allow Navies to interpret and act upon 'big data'. Navies generate massive amounts of data both organic and received; control and utilisation of the electromagnetic (EM) spectrum has long been recognized as a useful arena of war, in 1973 Adm Gorshkov of the Soviet Union is quoted 'The next war will be won by the side that best exploits the electromagnetic spectrum'. AI is the most advanced tool available to understand and consequently exploit the EM spectrum, it allows consideration of millions of data points against the entirety of human written doctrine and deliver insightful answers in an instant, within its own niche it can already outperform the human mind. Allowing Navies to grow confidence in algorithms' abilities to make decisions in place of humans, will lead to it steadily replacing them in rapid response decision making processes. It's not unfeasible to imagine a fully autonomous warship operating without humans onboard, at some stage in the next century. In the nearer term, AI is responsible for advancements in speech

[1] Project Renown: Improving Ships Availability – Jim Higham

recognition and data processing behind video streaming services. The 'Highly Autonomous Warship' (BMT - 2021) presents thought leadership on how warships could use AI.

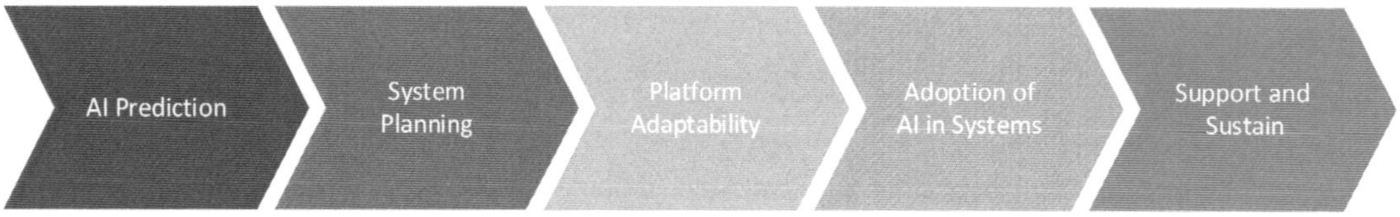

Figure 1 - AI Roadmap

From an obsolescence perspective it's important to understand how this technology might be embodied and sustained. As consideration of adoption of AI across all the seafaring and maritime warfighting disciplines develops, a view is needed on how AI should evolve, which systems could become fully automated, and which would derive the most benefit under the direction of an AI algorithm. Estimating both the size and technical system margins builds a picture to inform on the overall ship margin in respect to size, electrical load, and open architecture central computing power, making allowances for these needs at a suitable point in the life of the ship. This enables development towards making equipment adaptable within medium time frames (as defined by Moss and Kimber - 2019) having also allowed for sufficient margins to accommodate predictable AI adaptations. With AI adopted in ships systems, delivering tangible benefits, consideration is needed to ensure it remains useful and available with effective support mechanisms. These efforts ensure systems owners have more options to avoid obsolescence-driven, full system replacement and adapt what is already there.

2.1 AI PREDICTIONS

How AI develops is largely for Navies to decide. It requires funding and resource, therefore those areas best funded are likely to see the greatest advances. Predicting adversary AI capabilities will inform development to some extent, however academia may better understand the future potential of AI. General AI[2] is the academic goal, which, if even possible is still decades away, however fundamental rules frame the ethics of academic development towards it. Control of AI is about understanding intelligence, independent of the objective, and move toward systems which are provably beneficial to humans. To achieve this, there are three key rules to follow:

- *The machine's purpose is to maximize the realization of human values. In particular, it has no purpose of its own and no innate desire to protect itself.*
- *The machine is initially uncertain about what those human values are. The machine may learn more about human values as it goes along, of course, but it may never achieve complete certainty.*
- *Machines can learn about human values by observing the choices that we humans make.*

['Provably beneficial Artificial Intelligence', Stuart Russel]

Designers and system owners need to have a basic grasp of the AI rules to ensure they understand the direction in which AI systems are developing, by being better informed it's possible to better provide for future adaptability on board ships. To embrace change, the warship needs to be designed with AI in mind. For this to happen intelligently, designers and system owners need to ensure they are suitably educated in AI to imagine how, at the detailed level, AI might continue to prove itself beneficial to humans.

2.2 PLANNING FOR AI

The obsolescence management process should begin during the concept phase of procurement. Writing an Obsolescence Management Plan early in the life cycle forces consideration of how to manage obsolescence in an enduring manner.

[2] General AI is a term used to describe an AI algorithm capable solving almost all tasks a human can solve. Shevlin, H., Vold, K., Crosby, M., & Halina, M. (2019). The limits of machine intelligence: Despite progress in machine intelligence, artificial general intelligence is still a major challenge. *EMBO reports*, *20*(10), e49177. https://doi.org/10.15252/embr.201949177

Figure 2 describes the various ways in which obsolescence management should engage within the Concept, Acceptance, Design, Manufacture, In-service, Disposal (CADMID) structure.

<table>
<tr><th rowspan="7">Obsolescence Management</th><th>C</th><th>A</th><th>D</th><th>M</th><th>I</th><th>D</th></tr>
<tr><td>Is there an OM policy?

Have OM related user requirements been defined?</td><td></td><td></td><td></td><td></td><td></td></tr>
<tr><td colspan="2">Have design strategies to minimise obsolescence been identified with processes to ensure their use during design and design changes?</td><td></td><td></td><td></td><td></td></tr>
<tr><td></td><td colspan="2">Has an OM Strategy been developed?</td><td></td><td></td><td></td></tr>
<tr><td></td><td colspan="4">Has an OMP been developed?

Has an obsolescence risk assessment been conducted?

Have OM needs and tasks been flowed out and agreed across the organisation and stakeholders?</td><td></td></tr>
<tr><td></td><td></td><td colspan="3">Have OM requirements been flowed down to the supply chain?

Has a technical refresh plan been developed?

Is there a process to identify obsolescence resolutions as part of an options analysis?</td><td></td></tr>
<tr><td></td><td></td><td colspan="4">Has reclamation of equipment with useful remaining service life been included in the disposal plan for reuse or remanufacture?</td></tr>
</table>

Figure 2 – Obsolescence Management (OM) Through CADMID

To build a picture of what the future should hold, proactive engagement with warship design stakeholders is vital to plot the direction of development. Collaborative thought centres could provide the environment of innovation most likely to deliver provable benefits and identify methods of minimising obsolescence. There are several things that these working groups could consider to secure success in implementing and supporting autonomy in warships.
They should be asking questions pointed towards identifying future weaknesses and provoking AI-centred thinking, which obsolescence mitigation should have a stake in. To engender this thinking, working groups should consider the following:

- How can AI control this system in its current format, in a beneficial way? *What operations could be 'simply' interfaced with AI algorithm now?*

- How would AI control this system? *if the electromechanical interfacing was in place?*

- What space requirements might the system need? *How much space might the mechanical components require, and might it preclude human operation?*

- What computing needs might be required? *Use of centralised IT or pre-existing computing architecture?*

- Is the system small enough to incorporate change in a Fleet Time Support Period (FTSP)? *Consider zoning to create subsystem areas to enable incremental changes to take place within a FTSP.*

- In this highly autonomous warship what tasks are left for the human? *What technology is needed to make this task automated?*

- What steps should be taken now? T*o make this system ready for AI and the rapid changes in technology during its lifecycle.*

- How can industry create and sustain the plan to keep this technology up to date? *What technology refresh or updates are expected?*

These considerations should build towards making equipment, systems, and ships more autonomous, capable, and adaptable therefore less susceptible to show-stopping obsolescence.

2.3 ADAPTABILITY

Fully autonomous warships are not yet a viable reality, however a steady adoption allowing time to develop culture, technology, understanding, and doctrine surrounding the use of mechanical and decision-making automation is feasible. Adoption and normalization of automation will be incremental – as almost all large-scale decisions are taken with formal incrementation. It is of vital importance that all Defence Lines of Development (DLODs) are kept up to date with the technological advance to prevent it being unexpectedly derailed by a lack of consideration, from a non-technical viewpoint. The Defence AI strategy is due to be published later in 2022 by the Defence AI Centre, which will hopefully set out a near, medium and long term plan for AI adoption, which defence and its industries can use to inform their own development aims.

This process of incrementally introducing AI will need ships to be adaptable to change. The refresh cycle of this technology will likely be more frequent than the ships upkeep cycle, so by designing in adaptability, significant system updates should be incorporated into FTSPs to reduce the upkeep workload.

Designing for change at a system level should enable low level system components to be interchangeable with standard interfacing (where possible) and remain accessible. Implementing low level system updates reduces the potential for unnecessary component changes and therefore waste. This needs to be built into the ship at the system requirements level, as it will take the collective efforts all warship design stakeholders to develop a workable system.

3. ADOPTION OF AI

3.1 SOFTWARE AND INTEGRATION

The software integration process uses extensive land-based testing with facilities prior to use onboard. It is a safe space to integrate changes and deliver assurance that new technology is mature enough to enter service onboard. Using these facilities in a more agile context could help to reduce the timescales of incorporating AI and automation changes/updates. Greater application of the 'adaptability' mantra to software design will increase assurance risk; however, what's lost in terms of assurance could be mitigated by rapid adaptability paired with effective OEM collaborative support, thus reducing the impact of errors with timely resolution.

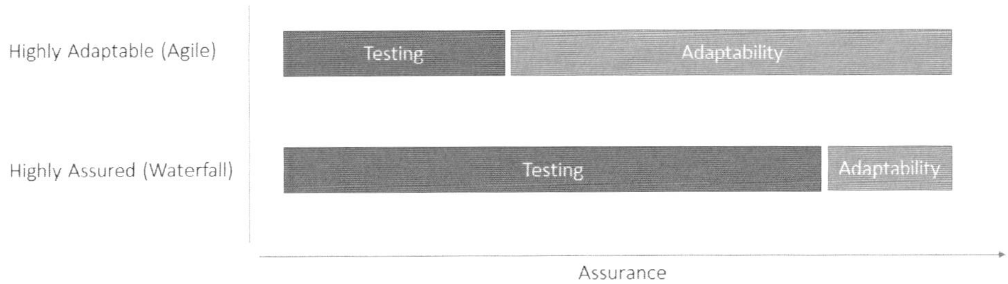

Figure 3 - Information Assurance Mitigation

The larger risk of implementing less tested software is mitigated by rapid adaptability, see Figure 1, meaning issues can be remedied quicker. On a case-by-case basis, the appetite for assurance risk will vary e.g. where risk to life is a major consideration of the software, significant testing will be needed to give high levels of assurance from the outset, with very low likelihood of failure. In-service software whilst in use will enjoy diminishing assurance risk over time, particularly with deep learning algorithms as its data set becomes larger and increasingly validated.

3.2 DOCTRINE AND LEARNING

Doctrine surrounding the use of AI is still in early development and needs to continue to prepare for coming change. Internally driven updates to AI algorithms with industry assistance must be central to the development doctrine, championing the core messages of the government guide to AI procurement[3]. These collaborative partnerships should have feedback loops to ensure that issues are remedied, and an Evolutionary Acquisition Lifecycle (EAL)[4] can be implemented; this is particularly crucial with the early adopters. Having a mix of experienced operators, keen developers, and empowered decision makers in the partnership to drive relevant changes will help to foster the EAL. The EAL is an acquisition cycle focused on uncertainty, and recognises that capability can be mature enough to acquire, but will need upgrades based on developing requirements; this evolution needs the right people involved.
Currently the majority of AI SQEP is held within industry, Navies should therefore consider widening the pool of their usual cadre to include software development SQEP amongst their ranks. The field of software development is specialist, and Navies will need more of these skills 'in-house' in the future. Improvement of AI comprehension across the organisation and developing AI doctrine will promote adoption and maximise its benefits. All this effort and change is to enable the organisation to prepare for future change, working with more diversity of thought and increased partnership to develop the most effective outcomes.

3.3 OBSOLESCENCE WITHIN AI

AI is fundamentally software, which can suffer from obsolescence. Resolving obsolescence in software requires a loop of considerations, reaching into both hardware and software, making its management complex. Interdependencies between hardware and software can create vast, constraining webs, causing complications when attempting to resolve obsolescence.

[3] Guidelines for AI procurement - GOV.UK (www.gov.uk)
[4] Evolutionary Lifecycle - Acquisition Lifecycles - KiD - UK MOD (mil.uk)

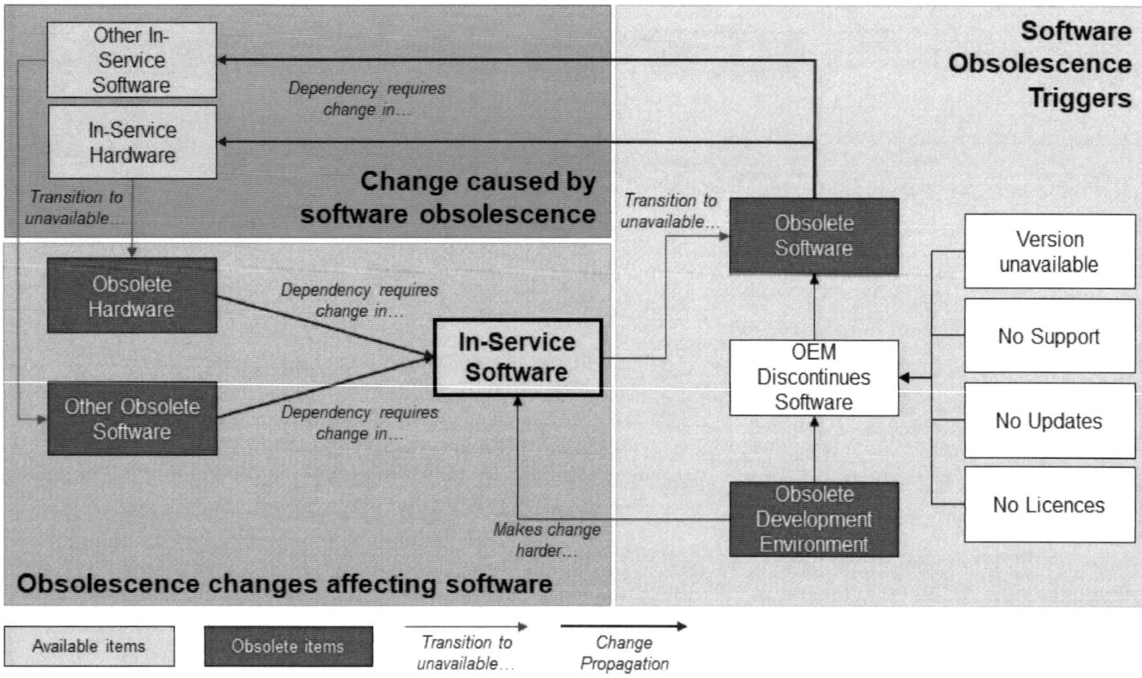

Figure 4 - Software Obsolescence Considerations

Software support is vital to the effective adoption of AI into Navies, this support needs to provide robust coverage across DLODs. It's a key component for the Navy to adapt and improve its digital skills base, taking a leading role in adapting technology for its own ends making it more aware and resilient to a loss of support.

AI over the medium term is going to be hardware specific, i.e. an algorithm designed for specific useful outcome within a defined system; therefore, the entire system needs to be considered from an obsolescence standpoint. Software obsolescence often results in software updates, which can have knock on impacts within the system, both in terms of other software, and hardware, see Figure 4.

Cascading change can result in vast projects to resolve obsolescence. In the case of the Type 45 Platform Management System (PMS) some aspects of hardware became impossible to procure spares, owing to the age of the system. The software required a large update to accommodate this new hardware, however, to achieve these changes, the software required more recent operating software. This new operating software wouldn't operate on the old servers, and thus the entire system was replaced. This story of knock-on impacts drove an initially simple update to vastly grow – because the aged system ran out of spare parts.

Software dependencies usually occur in user level applications as they typically rely on an operating system. Adding layers above this is to be avoided, generally. Coupled software describes the interdependency of applications; as such, where obsolescence affects one, its likely to impact the operation of the second. There are many types of software coupling, and if AI is tightly coupled to other software applications it's likely exposed to greater obsolescence risk. These interdependencies are described in Figure 4 as the transition from yellow to blue boxes.

In a militarily competitive environment, capability upgrades will drive development of AI even faster, with new generation algorithms likely to render existing iterations of adversary AI inferior. Constant improvements reduce the likelihood of this occurrence. In less competitive times, technical refresh planning, coupled with proactive obsolescence management reduces the likelihood of obsolescence impacting system availability.

4. SUPPORT AND SUSTAIN

To achieve MOD sustainability goals equipment needs to endure through change. Replacing entire systems may be commercially simpler to achieve, however the levels of waste generated conflict with the needs of the planet. Resilience can help towards achieving sustainability.

4.1 TECHNICAL REFRESH PLANNING

Where appropriate, when 'new to service' equipment is acquired, it should have an accompanying technical refresh plan, evidencing expectations of how supportability/obsolescence will impact both the software and equipment. Coupling technical refresh planning with a strategic AI plan will offer clearer direction of development for system owners. Modelling of how systems will evolve with AI insertion initiatives will rely on combining digital threads to understand it's future performance. This simulation, modelling and digital information management during development will aid the use of in-service digital twins to reinforce the benefits gained from the digitisation of engineering. Presentation of the possible technical direction that equipment could take enables equipment teams to adopt this thinking for real world use, and actively support its development. By presenting technical refresh plans in the design phase of the CADMID life cycle, ship delivery teams can financially prepare for the future; currently proactive equipment obsolescence horizons are only 10 years ahead.

4.2 RESILIENT THROUGH CHANGE

Obsolescence can occur in multiple ways. Preparing for it, having a plan, and proactive management helps towards mitigating risks. To minimise its cost to budgets and the environment, its best to achieve change at the smallest component level. This is true for both physical and digital projects provided they were designed for this type of low-level repair. AI, like many articles of software is vulnerable to obsolescence, and if designed or managed poorly, could result in the needless spend of vast amounts of effort on avoidable resolutions, with further consequences in the physical system.

There are multiple triggers of obsolescence risk in software as described in Figure 4, but loss of software support to an AI algorithm is probably the most disastrous, this risk requires mitigation. In this scenario the collaborative partnership between the Authority and Industry must be able to continue development of the algorithm, as without development it will become obsolete. To continue development, remaining members of the collaborative partnership must maintain access to, and control over the source code. Adequate provision of intellectual property rights must be organised to cover the eventuality of OEM support becoming unavailable, options include using 3rd party escrow or outright purchase. The partnership must have the legal authority to continue to improve and make software changes in the instance where the OEM is no longer a member of the partnership.

AI software used in the autonomous warship is valuable, and will appreciate, assuming an increased operational reliance and evolutionary improvement; it therefore needs adequate protection, because in time the very basic operation of ships will rely on it.

4.3 SUSTAINABILITY

Moving towards a more circular economy means making equipment adaptable, repairable and in the end recyclable. Material choice is important in this regard, as are design and commercial arrangements. Support contracts for AI systems should incorporate the opportunity for adaptation, repair, and improvement at the lowest component level, from both capability upgrade and obsolescence resolution perspectives. Full system replacement should be a last resort, to achieve maximum environmental benefit from the existing ship and its component systems. In today's linear economic model it's a tough challenge to balance the need to advance and improve, against the need to deliver long term, sustainable value from the equipment.

Confronting planned obsolescence will be a challenge; AI development should spell the end of 'black boxes' with heavily restricted options for adaptation – driven mostly by the commercial desire to force the customer into premature whole system replacement. Open architectures, protocols and interfaces enhance the ability to upgrade and replace components and articles of software, resulting in a reduced likelihood of full system replacement. In an arena of constant innovation, these systems must be designed to receive upgrades multiple times over their life. In a software context, layers of abstraction are needed to make the programming modular. With standard interfacing, modularity allows piecemeal upgrades and more isolated testing, accelerating the development cycle by virtue of reducing system interdependencies. In a hardware context, powerful, well maintained, and resilient servers are needed, with back up and redundancy available to multiple systems using an open systems architectures. Having a central machine with modular characteristics will enable replacement of individual component parts, reduce waste, improve repairability and contribute towards the strive for circularity.

5. CONCLUSION

Not keeping pace in the race to achieve computer-controlled automation will render the warship less able to respond to adversary AI threats. This Revolutionary development will change the battlespace, just as the jet fighter rendered human controlled anti-air defence useless, AI will eventually render human data assessment inferior.
Failure to ensure equipment adaptability, will result in greater barriers to change, owing to change being more intrusive and expensive, as the rate of technological development continues to accelerate. AI will have an increasing role to play in the control of equipment and it must remain adaptable to system change. Warships under design need to ensure they can adapt to this change, with reasonable measures sought to ensure adequate margin in design to accommodate future changes. Future system changes are likely to become more frequent, therefore systems should also be designed to accommodate changes within FTSPs.
Defence-wide education and collaboration in AI is important to ensure teams are aware of how AI will develop within their own sphere. AI is no longer the preserve of data architects and software engineers alone. Multiple, diverse viewpoints are needed to develop AI centred thinking in the right direction and achieve provable benefits.
Obsolescence, if not managed correctly will make adoption of AI incredibly costly both in terms of support and technical refresh. Obsolescence management and planning needs to begin at the outset of the equipment life cycle, and AI factors need to be considered for items of equipment potentially capable of receiving it in retrofit. AI is coming, now is the time to plan for it.

TOM FLINT holds the current position of senior consultant at BMT Defence and Security. He is responsible for providing obsolescence management advice and support to surface ship and submarine projects both in build and in service. His previous experience includes Company Director of an IT services company, providing software solutions for MOD, and as Weapon Engineer Officer in Royal Navy primarily in T23 Frigates.

SURVIVABILITY, ADAPTABILITY AND OFFBOARD SYSTEMS AUTONOMOUS AND OFFBOARD SYSTEMS AND THEIR IMPACT ON WARSHIP AND TASK GROUP SURVIVABILITY

D Manley FRINA, RCNC, University College London, UK

SUMMARY

Developments in Artificial Intelligence, autonomous navigation, tactics and doctrine, recruitment issues and a desire to reduce the combat risk to service personnel are combining to drive the rapid development of offboard capabilities, often autonomous in nature, in the maritime environment. The paper discusses the likely impact of this change in paradigm on naval survivability at platform and task force level before addressing developments in technology and process either in hand or required, that will enable these systems to reach their full potential whilst ensuring that current survivability and combat safety performance is not compromised.

1. INTRODUCTION

Recent years have seen a considerable interest in the development and deployment of offboard systems, either remotely controlled or autonomous in operation, in the maritime environment. Initial enthusiasm in events such as Unmanned Warrior have been spurred on by the success of autonomous systems in the Armenian-Azerbaijan conflict of 2020, the highly effective use of "Switchblade" and TB-2 Bayrakar drones against Russian forces on land and at sea in Ukraine and the successful use of waterborne and airborne drones by the Houthis in the Yemen and Saudi Arabian Peninsula, not least in repeated attacks against coastal infrastructure and the damaging or sinking of three warships in and around the Bab Al Mandeb Straits[1].

Figure 1 - A Houthi explosive boat in Yemen.

Despite a reputation for being somewhat reactionary in its outlook the Royal Navy has often been in the forefront of new and developing technology. Experiences over the last couple of decades though would suggest that these are often adopted and introduced to service before an understanding of the wider impacts of a new technology are recognised. This paper is

[1] Successful Houthi attacks against Coalition naval vessels using explosive boats include *HSV Swift* (October 2016), the Saudi frigate *Al Madinah* (January 2017) and the UAE minehunter *Al-Qasnah* (July 2017). Other attacks are known to have occurred.

offered as a starter for discussions on how the rise of offboard systems will impact the survivability of naval vessels, and some considerations for the design of those systems themselves.

In order to reduce the risk of inadvertent classification issues the discussion and examples are predominantly focussed on offboard systems deployed from or operating with surface ships. The principles however should also hold true for submarine-focussed systems, and it should be noted that there is considerable development activity in this area beyond commercial systems.

2. SURVIVABILITY

The word 'survivability' means different things to different people. In general use, survivability means the ability to survive a damage event to any degree a certain situation and most people are familiar with this definition. As a result, many in the naval community use this rather limited viewpoint when applying this to naval ships. However, in the naval context, survivability has a specific meaning, as well as a much broader application.

Allied Naval Engineering Publication 43 (Edition 2)[2] is NATO's handbook on ship combat survivability. It defines survivability as:

> *"the capability of a weapon system [or vessel] to continue to carry out its designated mission(s) in a combat threat environment".*

The element concerning the ability to continue the mission underlies a key aspect of maritime survivability. It is not sufficient for the vessel merely to remain afloat; to "survive" the vessel must strive to remain combat effective, and to be able to continue its mission.

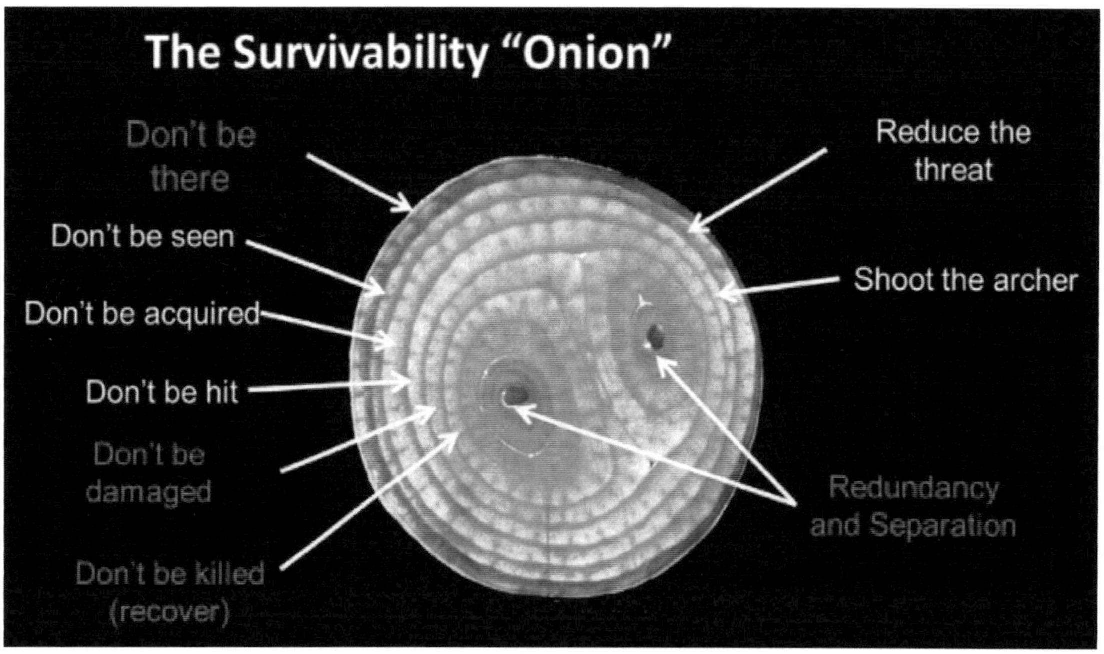

Figure 2. The Survivability "Onion"

Furthermore, survivability covers much more than just the design and construction of the vessel. Defensive systems, tactics, crew levels and training, the presence and capability of friendly forces, and the operating environment all play a key part in affecting the ability of a vessel to survive a threat and retain or recover her capabilities to be an effective member of the naval force. It is not just surviving a situation; it also considers how the vessel got into the situation in the first place, and how she will behave well after the situation is resolved.

[2] ANEP 43 is shortly to be published at Edition 3. The definition is likely to change in detail, emphasising the "platform" as well as the "system" but the sense of the definition will not be significantly altered.

Survivability comprises three primary elements, Susceptibility, Vulnerability and Recoverability[3].

- Susceptibility is concerned with the probability of being detected, targeted, engaged and hit by an incoming threat. It is commonly seen as being driven by the vessel's signatures. However, the reality is significantly more complex since, as well as signatures, it encompasses own end enemy force levels and capabilities, hard and soft kill systems on board ship and in the deployed force, rules of engagement, the environment, enemy and own crew training and a number of other aspects.

- Vulnerability (or, more frequently vulnerability reduction) looks at the short term effects of a successful weapon engagement. It is primarily concerned with the design of the vessel, the arrangement of and redundancy within its systems, armour and protective systems and automatic system reconfiguration. The key measure of effectiveness for vulnerability is the Probability of Escaping Critical Damage (P_{ECD}).

- Recoverability is concerned with the longer term spread of damage (for example, fire, smoke and flooding), damage control and firefighting and post-hit capability restoration. In extreme cases, where the vessel is not recoverable a transition is made into evacuation, escape and rescue. Hence there is a need for the vessel to be capable of acting as a safe haven for ships staff and a safe platform from which to evacuate should the need arise.

3. AN ANALOGY – AIRCRAFT AT SEA[4]

The development and deployment of offboard systems could be considered to be broadly analogous to the transition from battleship to aircraft carrier. Prior to the 1920s the primary source of naval power was the battleship. Heavily protected against shells and torpedoes, and armed with guns of up to 18", battleships had reigned supreme in the worlds battlefleets from the late 1800s. Aircraft began to go to sea in small numbers early in the Great War, often as seaplanes launched from converted ferries, or flying off platforms mounted on the main guns of battleships. The use of aircraft at sea was centred primarily on reconnaissance, although the ability to conduct air strikes form the sea was explored to a limited extent. One ship that would be recognisable as an aircraft carrier, HMS ARGUS, was completed by the end of the war but her use was limited and the capability in her aircraft not significant.

Carrier air power exploded in the 1930s and 1940s as the capability inherent in the aircraft available expanded. Increasing range allowed aircraft to act more effectively in the reconnaissance role, reaching out hundreds of miles. This in itself enhanced the survivability of host platforms and task groups, allowing detection and identification of threats at distance, and allowing commanders to decide to engage or, if faced with superior force, to take avoiding action early. But aircraft were now also able to act as effective strike aircraft, carrying bomb and torpedo loads that equalled or exceeded (depending on the number of aircraft engaged in an attack mission) the striking power of a battleship or battle squadron.

Ship design obviously had to change to accommodate significant numbers of aircraft, with large flight decks and hangar space required. This drove against the design principles established for the earlier battleships to a great extent – heavy armour was not practical, nor indeed was it required as the use of aircraft allowed strike at distances well beyond the range of a battleship's guns which were effectively rendered useless. However, designers and operators were still faced with the threat posed by the new capability that they themselves were deploying. It was reasonable to assume that, in combat against a peer adversary, if one's own aircraft were capable of striking effectively against an enemy's carriers that the reverse would be true – the enemy's air attack capability would be able to strike your own ships.

Carrier survivability then, as now, relied on a layered approach. The outer layer consisted of the carrier's own fighters providing a combat air patrol or launched on warning to engage enemy strike aircraft at distance. Aircraft that got through would face anti-aircraft fire from defending ships surrounding the carrier and the carrier's own weapons. But with reciprocity of effectiveness in mind the carriers themselves were expected to take hits and to survive, which meant that inherent survivability features were still required and tailored to the threat. Protection against bombs was problematical, solved in RN carriers through the use of armoured flight decks (which also proved their worth against Kamikaze aircraft – forerunners of the guided anti-ship missile, although in their case the guidance system was human). But armoured decks impinged on the height of hangars and reduced numbers and types of aircraft embarked, hence were one end of a scale of

[3] Some nations merge vulnerability and recoverability into a single element. Current UK practice is not to do this.

[4] It is assumed here that UXVs are being deployed from maritime platforms – ships or submarines. Of course there is the growing potential to deploy and/or control those systems from land or even the air, which presents a whole different but related series of survivability benefits and challenges.

compromise. Torpedoes were seen as the primary ship killing weapon (better to let in water to sink a ship than air), hence the side protection systems used in battleships were retained to a greater or lesser extent in all but the smallest carriers, their design and effectiveness increasing as trials, analysis and wartime experience showed what worked and what did not. Today the detail of the highly effective side protection systems used in US carriers is a closely guarded secret, but it is unlikely to differ radically from systems developed for battleships and carriers in the 1930s and 1940s.

Comparison with the current development and fielding of offboard systems is striking. For several years we have seen small scale deployment of small systems such as the MQ-8 Fire Scout and Boeing ScanEagle from existing ships[5], providing a limited but welcome adjunct to existing ISTAR capabilities. Underwater applications in the form of minehunting systems have been more extensive using systems such as the old PAP-104 and more recently the Seafox remote anti mine system supported by sensors mounted on the host platform and more recently on Underwater Uncrewed Vehicles (UUVs), but these systems are limited in terms of range.

However, the Royal Navy and other navies now finds itself on the cusp of an explosion of offboard systems of increasing endurance and capability. The need to embark, operate and maintain these offboard air, surface and subsurface vehicles in the numbers required to enable their effective use will drive a fundamental change in the design of the ships that carry them. So how will this new environment impact on the survivability of these new ships?

4. OFFBOARD SURVIVABILITY

Analysis and operational evidence has shown that there is a clear benefit to own force survivability (and the potential to increase lethality) across all domains from the deployment and use of remote, offboard systems. For example, the ability to detect threats at distance to allow early engagement or manoeuvring to avoid the threat. This characteristic of "distance" has led to a belief in some quarters that the ability to deploy offboard systems means the host platform will be stood back sufficiently from the threat, effectively nullifying that threat. If that were so, warships of the future could dispense with expensive survivability features, indeed dispense with expensive military features and become more "commercial" in their design. This is a persuasive argument and might well be true when facing a more conventionally designed enemy force that does not themselves enjoy the benefits of effective offboard systems.

Figure 3 - The MADFOX Automomous craft typifies current developments in the RN's developing portfolio of autonomous craft

[5] Drone ASW aircraft first went to sea in the 1960s with the failed DASH uncrewed anti submarine helicopter. Later the USN deployed reconnaissance drones from its reactivated *Iowa* class battleships during the First Gulf War.

However, against a peer threat the theory of reciprocity remains relevant – if one's own systems can be considered to be effective against the enemy then it is reasonable to effect similar levels of effectiveness against one's own forces. For example, if it is postulated that our UXVs[6] are stealthy and can detect enemy forces whilst remaining covert, or can penetrate an enemy's defensive screens then it is reasonable to assume that the drones of a peer threat can do the same. And if lethality through drone swarms is assumed to be a capability that our forces will enjoy then again it is reasonable to assume a threat swarm would be similarly effective. Thus, the need to design platforms and forces that are survivable remains, although the exact detail of the threat and the manner in which survivability is delivered is likely to alter. Indeed, if UXV capability becomes particularly advanced then the threat posed to our own host vessels is likely to be significantly enhanced unless our own capabilities are similarly developed.

As an example of UXV use, the outer layer defence of a ship or task group against anti ship missile attack may be augmented or provided by Uncrewed Aerial Vehicles (UAVs) and Uncrewed Combat Aerial Vehicles (UCAVs) carrying hard and soft kill systems, whilst other UAVs are used to detect and target hostile task groups for subsequent attack by UCAVs that are able to penetrate the enemy's defences. However, the reverse will also be true in a peer threat environment; our own UCAVs will have to face a capable offboard active and passive defensive screen, hence they will require to be stealthy to evade the defence, or carry their own active and passive defensive systems to penetrate the defence before they are able to engage the target. Similarly, if such a capability is something that we aspire to it is likely the opposition will also be fielding similar systems themselves. And open source evidence suggests that several nations are already well on the way to doing so.

Hence, if we believe that mounting an effective attack against an enemy host vessel or task group is possible then we must accept that the likelihood that enemy UXVs and weapons will penetrate our own outer layers causing a direct threat to the host platform cannot be discounted. Thus the need to maintain effective levels of inherent survivability in the host remains – signature control, hard and soft kill to reduce P_{EH}[7], vulnerability reduction featured to enhance P_{ECD}[8], effective damage control and repair facilities to enhance P_R[9].

This is relatively straightforward to manage at the platform/ship level from a vulnerability and recoverability perspective, since the manner in which threat weapons deployed from, or cued by, UXVs are unlikely to differ significantly from those experienced now – shock, blast, fragmentation being the prime examples[10]. Their effects on a ship are well understood, as are methods of analysis and mitigation. However, Susceptibility becomes a significantly more complex issue as it becomes increasingly reliant on the contribution of offboard systems to deliver effectiveness against the outer edges of the survivability "onion"; the survivability of those offboard systems themselves has to be considered in the overall equation since their ability to survive against a threat and to continue to deliver their defensive capability determines their contribution to the reduction in susceptibility of the host or task group; indeed one is likely to be faced with "onions within onions" as the survivability of the offboard system "screen" becomes increasingly significant.

The deployment of UXVs may also give rise to an unintended consequence if their signature is not sufficiently controlled. Soviet cruisers tasked with engaging and disabling NATO aircraft carriers were equipped with powerful long range air search radars, the theory being that a good way to detect the location of an aircraft carrier was to detect the point at which its aircraft appeared[11] and disappeared on radar as they launched and recovered. Similarly, during WW2 there were several occasions where carrier task groups were located by an enemy following aircraft back to their home carrier (and an occasion during the battle of Midway in 1942 where US strike aircraft located the Japanese Combined Fleet by following the course of a detached destroyer that was seeking to rejoin its place). Thus, UXV susceptibility has the ability to reveal the presence of an otherwise covert and stealthy host platform if its UXVs are easy to spot and track on their way to and from their host.

The manner in which UXVs are distributed in a task group may also impact on host platform survivability, especially if their carriage is concentrated in a few specialist vessels. In the same way as a WW2 aircraft carrier was a high value target so a modern UXV carrier is likely to be a primary target for an aggressor; destroy or disable the UXV carrier and the capability in the force is significantly degraded. This consideration may drive a distribution of task group UXV assets,

[6] UXV is a "catchall" term that encompasses uncrewed vehicles on, below or above the surface, in the sea, air or on land.

[7] P_{EH} – Probability of Escaping Hit
[8] P_{ECD} – Probability of Escaping Critical Damage
[9] P_R – Probability of Recovering from damage

[10] Whilst this is likely to be true in most cases there may be opportunities to deploy novel weapons from UXVs, for example conventionally generated EMP bursts, propeller/propulsor entangling systems to disable ships and submarines etc.

[11] This approach was also used by Argentinean radar operators in the Falklands campaign of 1982 to gain an indication as to where the UK's carrier task force was operating.

both in terms of embarkation and maintenance, but also command and control, across ships of a task group rather than concentrated in a few assets.[12]

5. ASSESSING PLATFORM AND FORCE EFFECTIVENESS

A possible way to approach this new susceptibility conundrum, at least for a single ship, is to consider the "Ship as a Task Group". As noted above, a modern UXV-enabled ship is analogous to an aircraft carrier. Host platform survivability will be impacted by the effectiveness of its offboard systems in detecting and defeating threats, but also by the ability of those systems to remain effective in a combat threat environment, i.e. their own survivability. In an environment where one posits one's own UCAVs delivering a strike capability at distance it is not unreasonable to assume a similar counter UAV capability being fielded in a UCAV.

Thus, ones' outer layer defending UCAVs (analogous to an aircraft carrier's combat air patrol) may well find itself engaged and defeated by threat counter UCAVs allowing a strike to follow through. Susceptibility therefore needs to take into account the effectiveness of ones' offboard systems to remain effective in the face of a threat, resulting in an analysis that looks near enough the same as current manned systems defending a carrier strike group, or ASW assets providing a defence against a submarine threat.

Fortunately, analysis tools exist which can address these kinds of scenarios, albeit currently focussed on traditional methods of warfare. Tools such as SADM for above water engagements and ODIN for underwater engagements could be readily adapted for use in the uncrewed environment, with a capping tool such as Dstl's MISSION providing the integration needed to provide an assessment of overall survivability. From this point it is logical to extend the scope of analysis from "ship as a task group" to a task group itself; the tools are the same, it is simply the scale of the analysis that changes.

This approach would also provide a mechanism to answer the ever-present question when considering requirements for UXVs – what is better, many cheap expendable UXVs or fewer more capable but expensive types? The answer to this question of course is "yes", since both options (and varying shades of grey in between) will be appropriate in particular scenarios. But at present the answers in even a few of these scenarios seem few and far between.

6. SURVIVABILITY OF UXVs

As well as challenging the consideration and assessment of host platform survivability the measures needed to address survivability in a UXV also present a number of issues and opportunities. There are essentially two categories of UXV from a survivability perspective:

- Cheap and disposable –operating in a low threat environment, not providing an essential capability or doing so whilst achieving survivability through sheer numbers
- Expensive and highly survivable – providing acritical capability and/or operating in a high threat environment where attrition is not practical;

Of course there is a "grey zone" between these extremes where increasing levels of survivability drive increasing operational benefit; the techniques discussed earlier in this paper would provide a framework to assist in determining appropriate levels

The scope of survivability measures that can be applied to a UXV will depend on the size and type of the vehicle. At a simple level one could consider the same techniques used in a traditional vehicle. In an uncrewed surface vehicle for example, shaping and emission control to reduce Radar Cross Section (RCS) and Infra Red (IR) signature could be applied. Removal of the need to embark inconvenient humans allows for smaller craft to be designed, instantly reducing their visual and other signatures. Vulnerability reduction features such as the incorporation of ballistic and blast protection, shock mounting and redundancy and separation of equipments (in a larger vehicle) to enhance the vehicle's resistance to the damaging effects of a weapon are obvious candidates for inclusion. Recoverability though has several challenges – but also benefits – arising from the lack of embarked personnel.

[12] Again, there are parallels with WW2 ship design and operation; The RN and IJN deployed seaplanes as search aircraft on cruisers and battleships as well as on aircraft carriers to enable task force situational awareness. In this case the primary driver was to maximise the number of strike aircraft on the carriers, but it also gave a considerable area search capability to those ships when operating independently or in task groups without carrier support.

The lack of personnel mean that any recoverability effort will have to rely on the inherent design of the vehicle and its internal systems. For example, firefighting and flood control would have to be managed using fixed systems, the vehicle itself could be designed with self-healing structure able to plug fragment and bullet holes through an inner layer that expands when heated or becoming wet[13]. Alternatively it may be that a UXV is designed "simply" to remain mobile and able to recover to home after damage, hence a reliance on inherent ruggedness and redundancy.

The lack of a crew means the ability of the UXV to reliably and effectively sense the amount of damage it has sustained and to report back (if controlled remotely), or to decide itself to withdraw (if working autonomously) is important. This is especially so where the UXV is deploying a sensor system since a damaged but otherwise working sensor could send back erroneous data (e.g. tracking or range errors), presenting an incorrect tactical picture if the error is not detected.

The lack of a crew does convey several recoverability design benefits, for example allowing the use of craft-wide inerting systems, either reacting to damage or inerted from the start), using materials that would otherwise be forbidden in crewed platforms, allowing for closed cell structures that would not be practical if crew movement was required. Choice of materials used in construction may be affected, for examples materials that are toxic or which release toxic gas if burned may be acceptable. This approach needs to be used with caution though since it would most likely impact on where and how UXVs could be stored; perhaps a more acceptable approach for UXVs operating in the maritime but from a land base.

Reliance on a vehicle's internal sensor network to detect and classify damage requires a step change from current sensor networks employed in ships, perhaps to a system similar in rigour to those used in space craft. For example, sensors themselves need to be survivable and effective through an incident to allow for evaluation of damage control effect. For example, it is easy to instal and operate a fire suppression system that initiates automatically, but very difficult to know when to switch a suppression system off. Most sensors (Cameras/heat detectors/thermal imaging) will be perfectly able to detect the onset of fire but are blinded by the post fire atmosphere (smoke, soot, heat) and many current types would be unlikely to survive the event. Currently there is a reliance on humans to re-enter the compartment and provide that analysis for effective decision making. Fixed damage control systems will also need to be themselves resistant to damage. Again, using fire as an example, extensive use is made of fixed firefighting systems in ships but there is an acceptance that weapon damage may disable or destroy those systems in areas where they are required; in this case ships staff can intervene using traditional methods, but this approach is not available to a UXV.

An effective damage control and monitoring system in a medium to large autonomous UXV is likely to require significant computing power, driving cost and introducing its own vulnerabilities. Distributed computing is likely to offer a solution to the latter issue, potentially driving the first. In a remote-controlled system that is monitored by human operators (or even an autonomous system that provides status updates to its operators but otherwise operates by itself) the need to provide effective reporting of damage, operation of recovery systems and recoverability effectiveness will be critical in reducing operator overload. Finally, sensor reliability will ned to be such that false alarms are reduced or eliminated, especially where an alarm would trigger a response in a fixed system – for example a firefighting system.

Another area of survivability that increases in importance from the lack of a crew is resistance to cyber attack and electronic warfare in a contested electromagnetic (EM) environment. For example, if the defence of a ship or a task group is reliant on offboard systems what happens if and when communications with those systems are disrupted? Potential adversaries have demonstrated some considerable skill in this area of warfare and it would seem unrealistic to assume that our offboard systems would be completely immune to this kind of attack[14]. Full autonomy allows for retention of capability within established mission parameters, but becomes problematic where the vehicle is required to deploy a weapon, since ethical self-limits preclude the shooting of weapons autonomously[15]. If the requirement for a "man in the loop" cannot be achieved due to failed communications between the UXV and the controller the capability is effectively lost. A similar concern relates to IXVs that go beyond the scope of a controller or other friendly forces to observe. The need to prevent interception / capture of a UXV, tagging to reveal the course of a UXV post-tampering and the revelation of his operating base or platform, or sabotage in the form of secreting an explosive device on board that detonates once the UXV has returned home present increasing challenge. As a side issue, the return of a UXV from "over the horizon", in particular an autonomous system, may present IFF challenges, especially as the numbers of systems increases.

[13] During WW2 wooden barges in the Far East were found to be difficult to sink with heavy machine gun fire since the fibres in the wooden skin of the barge expanded when wet, plugging bullet holes. Some barges survived thousands of hits from 0.3" and 0.5" machine gun rounds whereas steel craft would succumb to far fewer rounds.

[14] The supposed Iranian "hacking" and capture of a US RQ-170 drone in 2011 provides an example

[15] Autonomy as applied to weapon systems brings with it a series of moral and ethical questions. A long-held assumption, at least in Western militaries, is that whilst an autonomous system may be capable of making its own decisions over the majority of a mission, the final decision to release a weapon must reside in a human. Whether other nations observe such an approach is not certain.

7. CONCLUSION

Autonomous and offboard systems present many opportunities for enhanced operations on the maritime and other environments. To date neither are the full design implications of their integration into ships and submarines and the manner in which they contribute, both positively and potentially negatively to ship/submarine survivability and hence to combat safety and mission success is not clear. The force multiplier effect benefits which will undoubtably drive 'own' UXV introduction should not be an overriding determinant in forsaking platform survivability, which on the surface is an attractive proposition, since that case is not proven and is most likely not supportable in a peer threat environment.

Tools, techniques currently exist that can be adapted or evolved to enable informed decisions on survivability requirements for platforms and UXVs for their intended roles. Similarly, design tools and processes are available to deliver the likely survivability requirements for platforms and UXVs. Although there may be some inertia to develop and exploit survivability techniques to achieve demanding UXV survivability levels.

If efficient and effective use of these new capabilities is to be made then significant work is required to analyse and characterise platform and force survivability in an autonomous/offboard environment, the risks and benefits understood and appropriate decisions on the future course of ship and submarine design made. In recent decades several navies have adopted new technologies without developing an understanding of key aspects of the new capabilities and equipments that they have adopted, leading to potentially unpleasant discoveries downstream. The rise of offboard systems presents another opportunity for unpleasant surprises unless these fundamental aspects are investigated and addressed early.

8. AUTHOR'S BIOGRAPHY

David Manley is the MOD Professor of Naval Architecture at University College London, and Course Director for UCL's naval architecture MSc and the Submarine Design And Acquisition Course (SDAC). He is the MOD Deputy Technical Discipline Lead of Naval Architecture MOD, Senior Fellow for Maritime Vulnerability Reduction, and Head of Specialism for Platform Survivability. In this role he supports the UK's maritime enterprise in ship and submarine survivability, with involvement and engagement in all naval ship, submarine and weapon projects. His previous experience includes over 30 years working in UK naval safety and survivability, driving the development of the UK's maritime survivability strategy, combat safety and naval mission modularity, as well as supporting survivability development through NATO, ABCANZ and TTCP activities. He is a Fellow of the Royal Institution of Naval Architects and a Constructor Captain in the Royal Corps of Naval Constructors.

COMPREHENSIVE NUMERICAL HYDRODYNAMIC CAMPAIGN TO EVALUATE FRIGATE CONCEPT HULLFORM

F Gamboa, BAE Systems, UK

SUMMARY

A hullform was developed for a new frigate based on the parent hullform of an existing design. The new hullform is able to provide higher payload and stability performance, consequently increase its adaptability capability.
Initial bare hull resistance to the new hullform concept was estimated using CFD (Computational Fluid Dynamics).
After the final fairing, a hydrodynamic campaign was defined to measure the hydrodynamic characteristics of the new concept.
BAE Systems used, for the first time, numerical prediction methods to undertake a complete hydrodynamic campaign on a hullform. The campaign included Bare Hull Resistance, Propeller Nominal Wake Survey, Self- Propulsion, Appendage Alignment, Seakeeping and Manoeuvring. The current paper is focused on the Bare Hull Resistance Analysis.
A validation was made against the available experimental results for the parent hullform. The validation showed a good agreement. The validated numerical modelling was used in the new hullform evaluation.
The results showed that despite the larger displacement, the new hullform hydrodynamic performance is similar to the parent hullform.

NOMENCLATURE

ASW	Anti-Submarine Warfare
BOA	Beam Overall
CFD	Computational Fluid Dynamics
LOA	Length Overall
MRF	Moving Reference Frame
NCF	New Frigate Concept
PD	Parent Design
VOF	Volume of Fluid

1. INTRODUCTION

It was deliberated that future defence needs, meant that Frigates with an ASW (Anti-Submarine Warfare) role will carry more combat systems items at a higher position in the ship. However, in order to reduce design and build costs their size, structural design and shaftline components shall be similar to existing successful designs.

A direct consequence of this assumption is that the current frigates displacement and stability characteristics are not enough to accommodate all the future expectations from the costumers.

BAE Systems Naval Ships investigated if the hullform of stock design, which possesses exceptional hydrodynamic performance and has been optimised throughout its design process, could be modified in order to offer future costumers an increased combat systems capability without detriment to the hydrodynamic performance.
A New Frigate Concept (NFC) hullform was produced, based in the current design, which offered a higher displacement and stability at the same design waterline.

A complete hydrodynamic assessment was made in order to establish if the new hullform preserved the hydrodynamic performance of the parent design.

The hydrodynamic assessment was made using only numerical prediction methods. The engineering numerical techniques applied to the assessment were based on CFD (Computational Fluid Dynamics) calculations.

A validation exercise was carried out on the applied numerical models. The validation process consisted of comparing the numerical calculation results against the known experimental results of the parent design.

2. NEW FRIGATE CONCEPT - HULLFORM DEFINITION

The new frigate concept hullform was based on an existing stock design.
The design rational behind the new concept was to apply minimal modifications to the parent hullform that would generate an increase of the allowed displacement and enhance the stability characteristics.
The optimisation exercise required the modifications, to the parent design to be kept to a minimum in order to avoid noticeably compromising the hydrodynamic performance, to keep the shaftline train and finally allow to use directly the already existent structural arrangement of the parent design.
Considering the above, the new hullform has the following characteristics:

- Same LOA (Length Overall) and BOA (Beam Overall) of the parent design. Both dimensions are cost drivers, therefore to retain the cost similar to the parent design, these dimensions were kept;
- In order to increase displacement, the new design was allowed to increase its BWL (Waterline Beam). Furthermore, the transom was extended, the transom flap removed and a fledge integrated;
- The hullform definition in the forward area was kept, allowing the angle of half entrance and bow features such as the sonar dome to be retained;
- Bottom surfaces were kept the same as the parent design. Consequently all the shaftline components could be kept from the parent design.

The following figure illustrates the differences between the new hullform and the parent design.

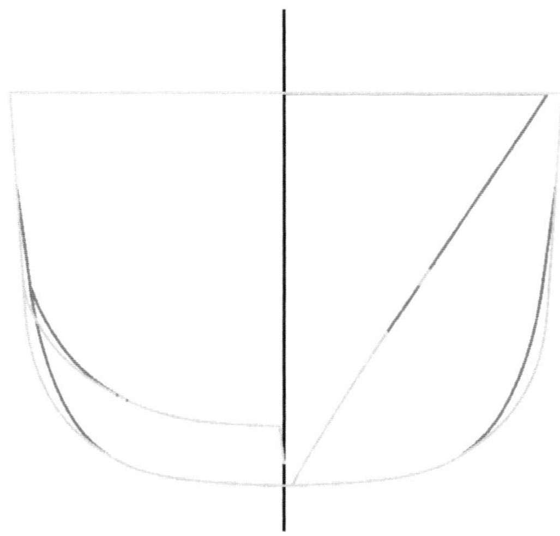

Figure 1. New Hullform Concept Sections (Green) and Parent Design Sections (Red).

3. HYDRODYNAMIC CAMPAIGN DEFINITION

To evaluate the full hydrodynamic performance the following topics were studied:

a. Bare Hull Resistance;
b. Nominal Wake;
c. Self–Propulsion and Propulsive Coefficients;
d. Seakeeping;
e. Manoeuvring.

The current paper will focus on the Bare Hull Resistance element of the hydrodynamic assessment:

3.1 NUMERICAL METHODS USED

The hydrodynamic campaign to evaluate the new hullform was made using only numerical prediction methods. The CFD package used was Star CCM+ Version 2020.2.

The following table summarises the numerical methods and the computational resources used in each one of the hydrodynamic assessments. Further details and modelling methodology are explained in the sections below. The nominal wake and self–propulsion elements are for reference only.

Table 1. Numerical Prediction Methods and Computational Resources.

Hydrodynamic Assessment	Numerical Method Used	CDF Package	Computational Resources [cores]
Bare Hull Resistance	CFD – VOF (Volume of Fluid) and MRF (Moving Reference Frame)	Siemens Star CCM+ 2020.2	124 Cores
Nominal Wake (for reference only)	CFD – VOF (Volume of Fluid) and MRF (Moving Reference Frame)		124 Cores
Self – Propulsion (for reference only)	CFD – VOF (Volume of Fluid) MRF (Moving Reference Frame) and Overset Mesh		256 Cores

3.2 NUMERICAL VALIDATION AND HULLFORM ASSESSMENT METHODOLOGY

The validation of the numerical simulations was made using the parent hullform that had been extensively model tested. To validate the CFD results, the model test runs were replicated using numerical prediction methods.
The validation methodology is the same for all the generated models.

It was anticipated that the use of similar discretisation and physical modelling for all numerical simulations would generate results affected by numerical errors and uncertainty of the same order of magnitude. This rationale allows us to directly evaluate the new hullform and the parent hullform by comparing directly the numerically predicted results.
This magnitude of the difference in the results was added to the known experimental result for the parent design to estimate the final result for the new hullform concept.

The following figure summarises the validation and comparison of the numerical results between both hullforms.

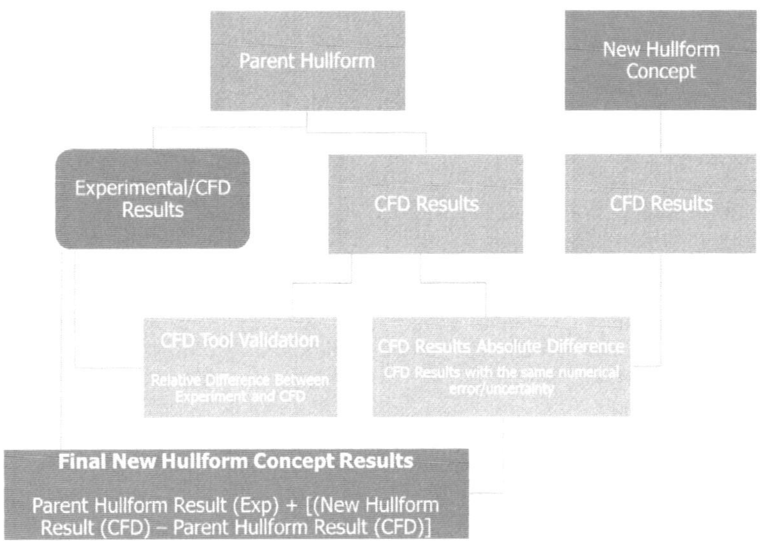

Figure 2. Validation and Hullform Assessment Methodology.

4. BARE HULL RESISTANCE

In this section the CFD simulations methodology, used for the validation and calculation of the New Hullform Concept bare hull resistance, are explained.
The methodology used is the same for all the generated models. The only parameters that were altered, depending on the simulation, were the following:

- Hullform;
- Draught;
- Displacement;
- Longitudinal Centre of Buoyancy;
- Ship Speed.

Siemens Star CCM+ version 2020.2 was the CFD package used throughout all the analysis.

4.1 NUMERICAL PREDICTION USING CFD – DISCRETISATION MODELS

All the CFD simulations were made with high fidelity models with the following mesh characteristics:

- Simulation at full scale;
- Initial Level Trim or Bow Trim;
- Refined mesh in areas where the flow is expected to be more turbulent or where there are larger speed gradients, for example bow and fledge areas in order to capture breaking waves and transom wave system;
- The bare hull contains skeg and sonar dome as per the tank test experiments;
- Kelvin wedge and free surface mesh refinements were used, the Kelvin wedge refinements intention is to capture the transversal and divergent wave system created by the ship whilst at speed;
- The boundary layer was captured using prism layers next to the hull surface. The aim is to keep the average y+ value around 100;
- Trimmed Cell Mesh, these orthogonal cells are aligned to the incoming flow and allow for a better free surface definition.

The resulting mesh had a high resolution with approximately 6 million cells for each simulation.

4.2 NUMERICAL PREDICTION USING CFD – PHYSICAL MODELS

The physical models used in the simulations were chosen to best represent the real aspects of the flow and follow the best practice guidelines from the CFD package supplier, the know-how acquired during previous projects and finally the experience acquired from the participation in Joint Investigation Projects (JIPs).
The physical models used are summarized below:

- VOF (Volume of Fluid) – Models the free surface and the interaction between air and water;
- Rigid Body Motion – Ship free to sink and trim dynamically during the simulation until an equilibrium is reached;
- Moving Reference Frame – The "laboratory" reference of frame moves to keep the free surface matching the sink and trim of the ship;
- $K - \varepsilon$ Turbulence modelling;
- Symmetry used – Only half of the hull modelled – higher mesh fidelity;
- Full Scale Simulation;
- Simulated time of 500 seconds.

4.3 ANALYSED RESULTS

The following data was calculated for each simulation:

- Bare hull Resistance;
- Dynamic Sinkage;
- Dynamic Trim;
- Free surface elevation;
- Pressure field in the hull.

The average results of the last 200 seconds of simulation were used as the final calculated result.

The standard deviation of the Bare Hull Resistance, Dynamic Sinkage and Dynamic Trim, for this period of time was also calculated to assure the convergence of numerical simulation. If the standard deviation of the last results was greater than 1% of the average results, the simulation would be run for a longer period of time until convergence was guaranteed.
In the validation exercise, the final numerically predicted results were compared against the available experimental results.

4.4 VALIDATION - CALCULATION MATRIX

To validate the CFD results, the model tests runs were replicated using CFD simulations.
Therefore, the validation CFD numerical simulations was made using the Parent Hullform which has a transom flap. The new hullform design has a fledge with the same angle of attack as the Parent Hullform transom flap.
In summary the validation exercise consisted of the direct comparison of the averaged CFD numerical results, for the last 200 seconds of simulation, against the equivalent experimental results.
The results that were compared are the following:

- Bare hull Resistance;
- Dynamic Sinkage;
- Dynamic Trim.

The experimental results at full scale are calculated using Froude scaling from model scale. The CFD numerical results were obtained directly from full scale simulation. Due to this fact, it was expected that the lift of the transom flap would not be correctly scaled because it is mainly Reynolds, not Froude number, dependent.
Therefore, there was expected to be a deviation of the results at faster speeds, where the transom flap produces more lift. This deviation was observed in other experiments [1].
In order to validate the CFD capability to predict the trim impact, at two displacements, CFD simulations were made at level and bow trim.
The following table, shows the CFD numerical simulations for the validation exercise.

Table 1. Bare Hull Resistance Validation Calculation Matrix

Hull	**Parent Design Hullform**					
Trim	**Level Trim**	**Level Trim**	**Level Trim**	**Bow Trim**	**Level Trim**	**Bow Trim**
Loading Condition	**Ballast**	**Design**	**Deep**	**Deep**	**Extreme**	**Extreme**
Tested Speeds	12	12	12	12	12	9
Total Validation Runs	69					

4.5 NEW HULLFORM EVAUATION - CALCULATION MATRIX

The CFD modelling methodology proposed to calculate the New Frigate Concept Hullform Bare Hull Resistance was similar to the validation runs.
The bare hull resistance for the four draughts at level trim, that were experimentally tested, was calculated using CFD. The use of the draughts as the point of comparison, instead of the displacements, is due to the fact that the New Hullform has a wider waterline beam and larger underwater hull sections, therefore, a higher displacement at the same draught than the Parent Design Hullform.
Furthermore, the impact of the trim in the resistance was calculated at two different mean draughts.
The following data was calculated for each run for the 500s of simulation:

- Bare hull Resistance;
- Dynamic Sinkage;
- Dynamic Trim;
- Free surface elevation;
- Pressure field in the hull.

The average results of the last 200 seconds of simulation was used as the final calculated result and compared against the experimental results and numerical results of the validation exercise.
The following table shows the CFD numerical simulations that were modelled and run.

Table 2. New Frigate Concept Bare Hull Resistance Calculation Matrix

Hull	**New Frigate Concept**					
Trim	**Level Trim**	**Level Trim**	**Level Trim**	**Bow Trim**	**Level Trim**	**Bow Trim**
Loading Condition	**Ballast**	**Design**	**Deep**	**Deep**	**Extreme**	**Extreme**
Tested Speeds	12	12	12	12	12	9
Total Validation Runs	69					

4.6 CFD VALIDATION - RESULTS

As mentioned before, the validation exercise consisted of the direct comparison of the averaged CFD numerical results, for the last 200 seconds of simulation, against the equivalent experimental results.

The following table and figures show the calculated results for the bare hull resistances.

Table 3. Parent Design Experimental vs. CFD Results – Relative Differences

Hull	**Parent Design**					
Trim	**Level Trim**	**Level Trim**	**Level Trim**	**Level Trim**	**Bow Trim**	**Bow Trim**
Loading Condition	**Ballast**	**Design**	**Deep**	**Extreme**	**Deep**	**Extreme**
Speed	**Relative Difference [%]**	**Relative Difference [%]**	**Relative Difference [%]**	**Relative Difference [%]**	**Relative Difference [%]**	**Relative Difference [%]**
0.21	3.45%	-1.23%	-0.45%	-3.81%	-6.25%	5.25%
0.29	1.20%	-2.91%	0.50%	-6.14%	0.21%	0.13%
0.36	-0.28%	1.09%	0.71%	-1.35%	1.83%	-2.38%
0.43	0.08%	1.43%	-0.83%	0.32%	1.83%	-0.60%
0.50	1.54%	-0.21%	-0.15%	1.48%	0.78%	0.06%
0.57	-0.98%	-2.43%	-3.32%	-3.30%	-1.31%	-2.79%
0.64	-0.34%	-1.16%	-4.50%	-6.51%	-3.97%	-2.86%
0.71	2.16%	-4.36%	-1.57%	-3.35%	-6.99%	-2.96%
0.79	6.15%	1.85%	-0.35%	-4.82%	-0.19%	-4.11%
0.86	8.38%	7.58%	6.62%	4.97%	6.60%	5.42%
0.93	7.67%	7.99%	8.52%	8.82%	7.58%	8.71%
1.00	6.66%	7.07%	7.54%	8.64%	6.89%	7.61%

In the table above when the relative difference is positive, the CFD bare hull resistance has a smaller value than the correspondent experimental value and vice-versa.

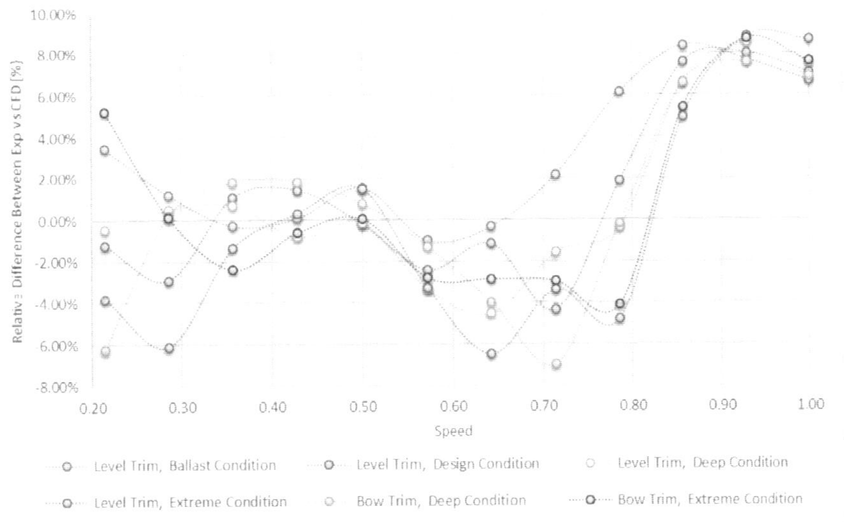

Figure 3. Parent Design Experimental vs. CFD Results – Relative Differences.

For practically all the simulations performed, it was observed that the absolute difference between the CFD predicted sinkage and trim against the experimental values was very small. Due to the insignificant difference that was observed throughout all the analysed loading conditions at different speeds the results won't be shown in the current paper.
Nevertheless, it is noted that the insignificant difference between the predicted sinkage and trim versus the experimental data reflects that the CFD methodology can accurately predict these parameters.

Focusing on the bare hull resistance validation, in Table 3, the relative difference between the experimental data and the CFD bare hull results is revealed.
Analysing the relative differences it is possible to identify two speed ranges where the dissimilarities between CFD and experimental results are greater: from 0.21 to 0.29 and from 0.64 to the top analysed speed of 1.

For the speed range between 0.21 and 0.29, the relative differences varies from 5.25% to -6.25%. The average relative difference is -0.84% amongst all the analysed conditions for the same speed interval.
Such a spread of results can be explained by the experimental error associated to the measurement of small resistance forces at low speeds. The tank test facility has commented that below a speed of 0.36 the reliability of the results is more difficult to guarantee. CFD can take measurements of any order of magnitude combined with the fact that the resistances predicted by CFD are at full scale, one can argue that CFD is more reliable for the calculation of the bare hull resistance at smaller speeds than acquiring experimental data.

Analysing the speeds between 0.36 and 0.57, the relative differences varies from -3.32% to 1.83%. The average relative difference is -0.37% amongst all the analysed conditions for the same speed interval.
In this speed range the mismatch is mostly within the expected experimental error of 2%, confirming the ability of the CFD methodology to predict the bare hull resistance. Further to the experimental error, some part of the relative difference can be credited to the Froude Scaling used to scale the model test results to full scale predictions.
CFD results were calculated directly at full scale, however the experimental values were calculated using Froude scaling. Froude scaling can add scaling errors if certain features of the flow are not Froude dependent but Reynolds number dependent.
Froude scaling also does not account for reattachment of the boundary layer that can happen in model scaling, however such is rare due to the use of turbulence stimulators in the model.

At high speeds, from 0.64 to the top analysed speed of 1, the mismatch between CFD results and experimental data for the bare hull resistance is the greatest observed in this validation exercise.
At this speed interval, the relative differences varies from -6.99% to 8.82%. The average relative difference is 2.65% amongst all the analysed conditions.
It is observed from Figure 3 that at this speed interval the relative difference climbs gradually and consistently for all the loading conditions analysed to values between 4.97%-8.82% at the top speeds of 0.86, 0.93 and 1.

The fact that the climb is gradual and occurs consistently at all the loading conditions point out that the CFD results show a full scale phenomenon that was not predicted by the model scale experimental results and corresponding Froude scaling to full scale results.

One plausible hypotheses is the effect of the transom flap at full scale. From the speed of 0.57 to the top speed the flow is detached from the transom flap, that is, the transom wave is not touching the ship's transom. Such is demonstrated in Figure 4.

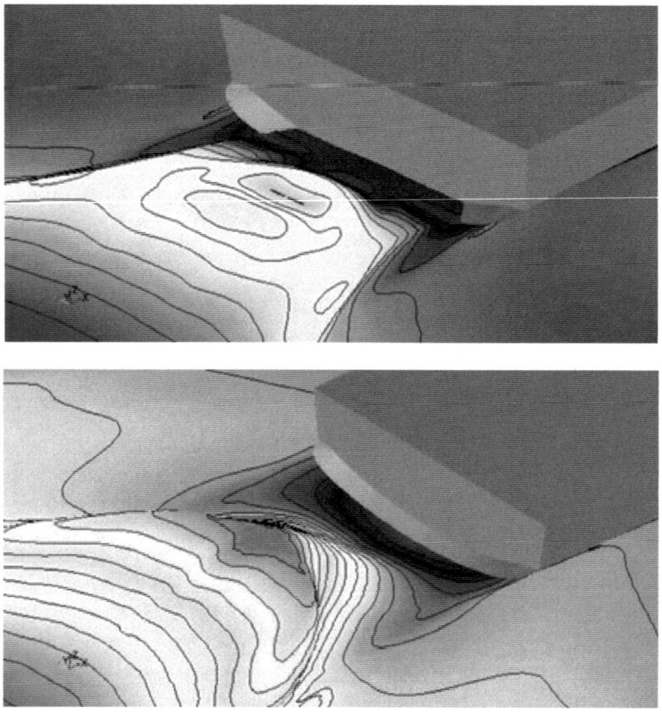

Figure 4. Transom Flap Attached Flow (Top), Transom Flap Detached Flow (Bottom).

The lighter the displacement, the slower is the speed which the transom wake detaches from the transom.

At higher speeds, the transom flap lift is also higher. The flow in the transom flap is Reynolds dependent and not Froude dependent. Therefore, it is expected that there are strong scaling effects at higher speeds, where the transom flap has more influence in the bare hull resistance.

The effect of the transom flap scaling is well documented in [1], in this paper a destroyer type of hullform fitted with a transom flap is analysed at 3 different model scales and at full scale. The study shows that even at model scale there is already a noticeable a difference due to the flap scaling effects. The difference accentuates when compared to the full sea trials as seen in the current validation exercise.

The final step of the validation exercise was to confirm that the trim effect in the bare hull resistance was captured by the CFD calculations.

The following tabulated results and graphs demonstrate that the trim effect on the bare hull resistance is very well captured by the CFD simulations.

Table 4. Bare Hull Resistance, Trim Effect – Validation Exercise

Testing	**Experimental**	**CFD**	**Experimental**	**CFD**
Trim Effect	**Relative Difference to Level Trim Results (+ve decrease in resistance compare to level trim results)**		**Relative Difference to Level Trim Results (+ve decrease in resistance compare to level trim results)**	
Loading Condition	**Deep Condition**		**Extreme Condition**	
Speed	**[%]**	**[%]**	**[%]**	**[%]**
0.21	11%	5%	-2%	7%
0.29	6%	6%	-1%	5%
0.36	4%	5%	6%	5%
0.43	2%	5%	5%	4%
0.50	5%	6%	6%	5%
0.57	4%	6%	4%	5%
0.64	3%	4%	4%	7%
0.71	5%	-1%	5%	5%
0.79	2%	2%	3%	4%
0.86	3%	3%	3%	3%
0.93	3%	2%	3%	3%
1.00	2%	1%	4%	3%

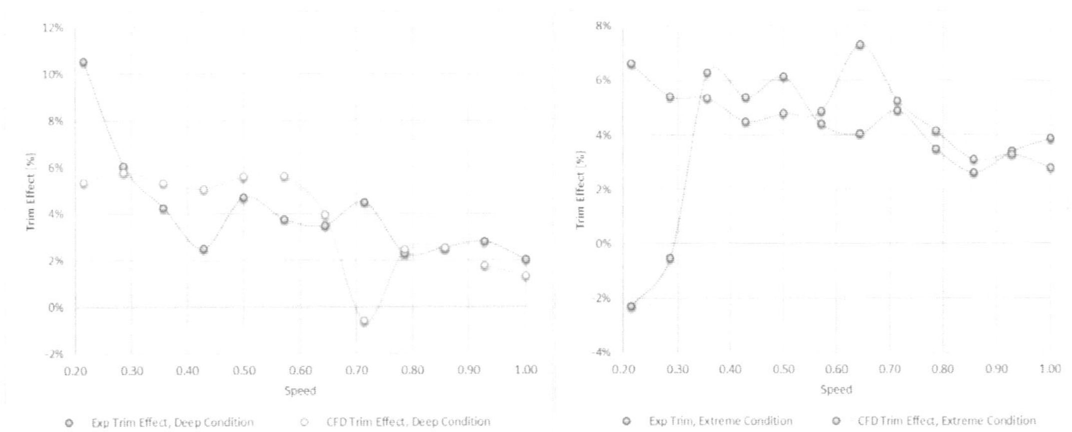

Figure 5. – Bare Hull Resistance Trim Effect – Validation Exercise

4.7 CFD BARE HULL RESISTANCE, NEW FRIGATE CONCEPT VS. PARENT DESIGN

In this section a direct comparison of the averaged CFD numerical results, for the last 200 seconds of simulation, is made for the NFC Hullform against the Parent Design Hullform.

The following table and figure show the calculated results for the bare hull resistance.

Table 5. Relative Difference between NFC Hullform and Parent Design Hullform, CFD calculated Bare Hull Resistance

Parent Design vs. NFC Hullform, Bare Hull Resistance, CFD Results – Relative Differences						
Trim	**Level Trim**	**Level Trim**	**Level Trim**	**Level Trim**	**Bow Trim**	**Bow Trim**
Loading Condition	**Ballast Condition**	**Design Condition**	**Deep Condition**	**Extreme Condition**	**Deep Condition**	**Extreme Condition**
Speed	**Relative Difference [%]**	**Relative Difference [%]**	**Relative Difference [%]**	**Relative Difference [%]**	**Relative Difference [%]**	**Relative Difference [%]**
0.21	3.50%	2.57%	1.00%	-0.55%	1.56%	1.40%
0.29	2.15%	1.70%	0.67%	-0.23%	1.24%	0.34%
0.36	1.76%	0.85%	0.15%	-0.33%	0.84%	0.13%
0.43	1.49%	0.76%	-0.12%	-0.64%	0.55%	-0.25%
0.50	1.02%	-1.23%	-1.52%	-4.30%	-0.15%	0.06%
0.57	1.62%	-0.33%	0.07%	-4.30%	0.65%	0.21%
0.64	3.39%	1.33%	1.61%	-1.56%	0.71%	2.12%
0.71	1.53%	-2.79%	1.25%	-0.19%	-2.45%	1.01%
0.79	0.81%	1.35%	2.41%	-5.93%	0.73%	1.80%
0.86	1.22%	0.97%	1.54%	2.55%	0.81%	1.70%
0.93	2.13%	2.31%	2.98%	3.57%	1.98%	2.77%
1.00	3.58%	3.66%	4.23%	4.38%	2.73%	4.06%

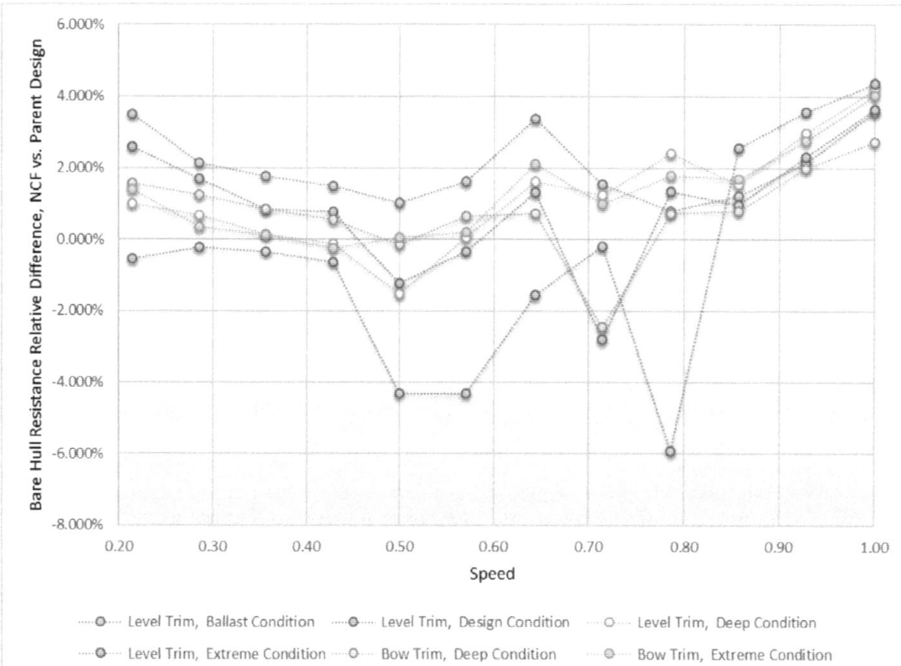

Figure 6. – Relative Difference between NFC Hullform and Parent Design Hullform, CFD calculated Bare Hull Resistance

For most of the simulations, it was observed that the absolute difference between the Parent Design and NCF Hullform CFD predicted sinkage and trim values was very small.

Due to the fact that both hullforms simulations were modelled using the same CFD methodology described in 2.1 and 2.2, it is assumed that all the results are affected in the same way by any discretisation errors and uncertainty.

Therefore, it is assumed that the absolute differences observed in the CFD values of both hullforms would be the same differences if both hullforms were tested experimentally.

In Table 5, the relative difference between both hullforms CFD bare hull results is revealed. The relative difference is calculated using the Parent Hullform result as the reference result.
Therefore, when the relative difference is positive, the Parent Design CFD bare hull resistance has a smaller value than the correspondent NFC Hullform value and vice-versa.

Due to its wider beam and higher wetted surface area, it is expected that the NCF bare hull resistance is also higher than the Parent Design in all the analysed loading conditions and speeds.
The higher resistance is noticeable in the average relative difference between both hull resistances. For all the loading conditions analysed and all the speeds, the NCF Hullform has an average bare hull resistance 0.93% higher than the PD Hullform.

The resistance difference is small at lower speeds between 0.21 and 0.43. In this speed interval, the relative differences varies from -0.64% to 3.50%. The average relative difference is 0.86% amongst all the analysed conditions.
Contrary to what was observed at lower speeds, the difference is no longer insignificant at top speeds.
The higher waterline beam over length ratio of the NCF Hullform has an impact in the residuary and mostly important in the wave resistance. The wave resistance component is the main resistance component at these high speeds.
In the speed interval between 0.86 and 1, the relative differences varies from 0.81% to 4.38%. The average relative difference is 2.62% amongst all the analysed conditions.

These results confirm the expected result of a slightly higher bare hull resistance of the NCF Hullform when compared directly against the PD Hullform.
Nonetheless, there are odd results in the middle range of the analysed speeds that show a lower bare hull resistance for the NCF Hullform.

With a deeper analysis, it's possible to observe that at level trim for the extreme loading condition, at the speeds of 0.5, 0.57 and 0.79 (Table 5, values in green), the NCF Hullform shows difference in the bare hull resistance that varies between -4.30% to -5.92%.
The hypothesis for this discrepancy is based on the impact of the fledge and transom extension that is part of the NCF Hullform design in the transom wake and the transversal wake just aft of the transom.
From the post processing of the CFD results it was observed that the pressure component of the resistance is where the difference is notable at these speeds. The transom flap of the PD Hullform introduces a lower pressure region just aft of the flap that accentuates the transverse wave and increases the pressure resistance. The transom fledge produces a smoother transom wake. This is shown in Figure 8, where the transom wakes of the Parent Design and NCF Hullform are directly compared.

A final analysis in the transom flow at the speed of 0.79, where the relative difference show the lowest bare hull resistance calculated, shows that not only the transom wake is smoother due to the presence of a more favourable pressure gradient, but also the general divergent wave system is smaller.
It is suggested that there is at the deepest draught analysed, a positive interaction between the bow wave and shoulder wave of the NFC hullform decreases in general the divergent wave system. The positive interaction could only come from the wider hullform in the shoulder that creates a higher pressure area, that consequently creates a different wave system than the one observed for the Parent Design.
The hypotheses above are demonstrated by the comparison between transom wake and wave system images for both hullforms:

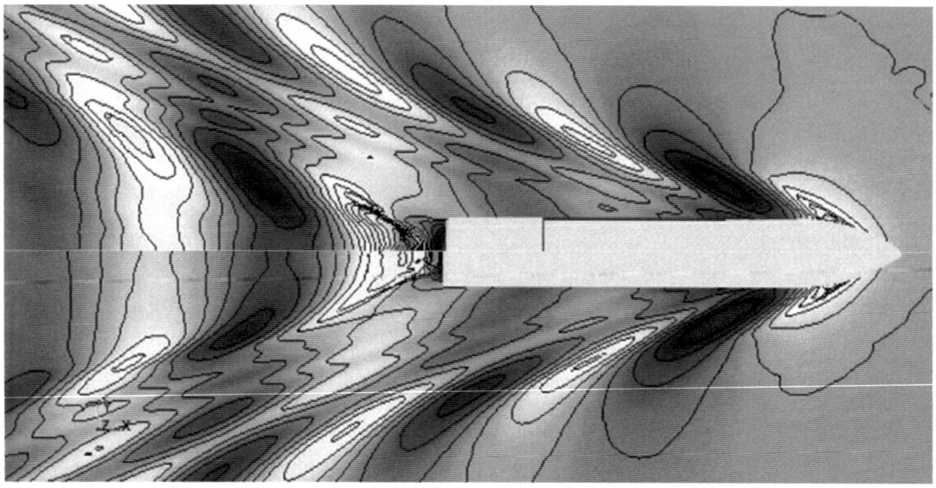

Figure 7. – Wave System at High Speed. Top Half: Parent Design Hullform, Bottom Half: NFC Hullform

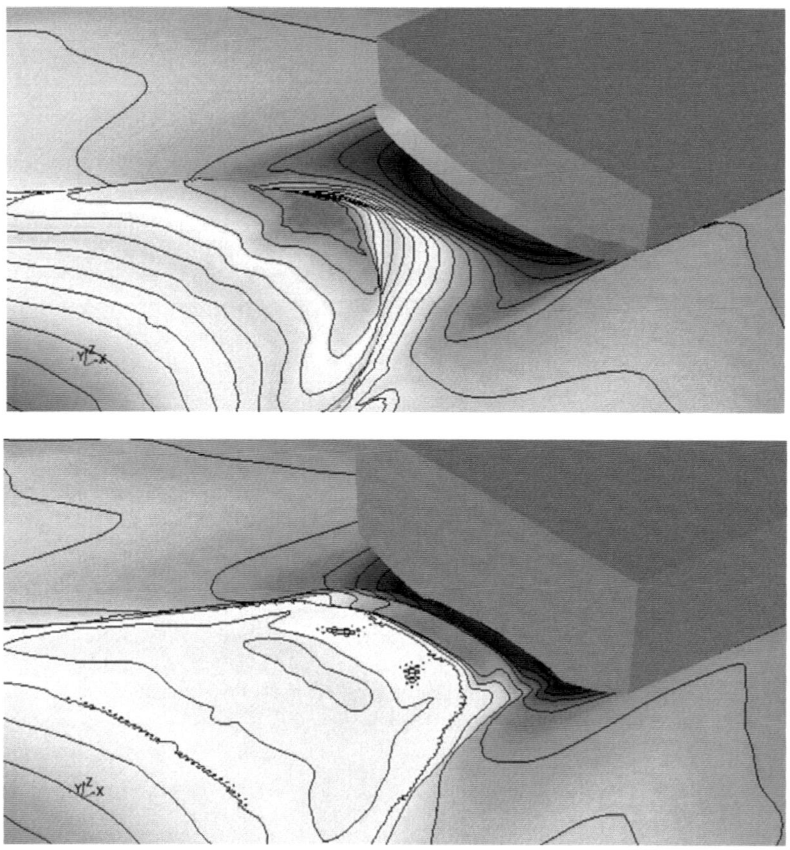

Figure 8. – Transom Wake at High Speed. Top: Parent Design Hullform, Bottom: NCF Hullform

5. CONCLUSIONS

A New Concept Frigate (NCF) Hullform was developed based on an existing Parent Design Hullform. The NCF hullform provides greater buoyancy and stability reserve, making it more attractive to potential costumers aiming to integrate extra combat systems items in the vessel.

The NCF waterline beam is wider than the Parent Design, and a detriment in the hydrodynamic performance in terms of bare hull resistance was expected. The NCF hullform design includes a transom extension and a transom fledge.
A CFD (Computational Fluid Dynamics) study was carried out to assess the NCF hullform bare hull resistance against the Parent Design Hullform.

The study included a CFD tool validation against the available Parent Design experimental results, and finally a bare hull resistance delta calculation between the NCF and Parent Design Hullforms.
The CFD methodology used throughout the study was confirmed by the validation exercise. The comparison between experimental and numerically predicted bare hull resistance for the Parent Design Hullform established a good correlation apart at lower and top speeds.

The mismatch between experimental and numerically predicted results at lower speeds are explained by the experimental uncertainty being greater at low speeds and finally the mismatch at high speeds was already expected and previously documented by [1], where scaling effects are experienced by the transom flap.
To complete the validation exercise, the effect of the ship's trim in the resistance was tested. The effect observed in the experimental data matched the one observed in the CFD calculations.

Using the validated CFD methodology a numerical CFD comparison was made between the Parent Design and the NCF Hullform. The observed results confirm the expectation that the NCF Hullform has a slightly higher bare hull resistance when compared to the Parent Design. The difference is greater at higher speeds. This is explained by the expected increase in the wave resistance due to the wider beam. Nevertheless, some results shown the opposite trend. A further analysis shown that the transom fledge has a beneficial effect at the higher displacement, smoothing the transom wake compared to the transom flap. Another added effect is a positive interaction between the bow wave and the forward shoulder wave that reduces the overall ship's divergent wave system at the higher displacement analysed.

Overall, the results showed that despite the larger displacement, the new hullform hydrodynamic performance does not show a significant detriment when compared to the parent hullform. The benefits of the increase in payload and stability characteristics of the NCF hullform surpass the slight bare hull resistance increase of 0.93%, in average, across all speeds and loading conditions analysed.

The calculated resistance difference between the CFD results of the Parent Design and the NCF Hullform was added to the Parent Design known experimental results in order to calculate the final NCF Hullform bare hull resistance.
The final outcome of this study was the NCF Hullform bare hull resistance curves at six different loading conditions, four at level trim and two at bow trim.

6. ACKNOWLEDGEMENTS

I wish to show my appreciation to Richard Trumper and Nick Pattison to sponsor the acquisition of the CFD capability to BAE Systems Naval Ships and believe in the CFD potential use for current and future projects.
I extend my appreciation to Gavin Rudgley, Stuart Hunt and Neil Griffiths who supported the submission of this paper allowing the time to gather the necessary information and write it up.
Finally, my special appreciation to Gary Webb who has been my mentor in all the hydrodynamic and sailing yacht matters since I joined BAE Systems Naval Ships.

7. REFERENCES

1. Naval Surface Warfare Centre Carderock Division, 'Scaling Effects on Stern Flap Performance Progress Report', *NSWCCD-50-TR-2009/079*, 2009

8. AUTHORS BIOGRAPHY

Fernando Gamboa holds the current position of Consultant Naval Architect, Hydrodynamics at BAE Systems, Naval Ships. He is responsible for surface ships hull design and hydrodynamic analysis of current and new projects. His previous experience includes CFD analysis and hydrodynamic performance evaluation for power and sailing yachts.

Warship 2022: Designing Warships for the Future
8th-9th of June 2022
M Shed, Bristol, UK

RESOLVING THE IMPACT OF SHIP'S DESIGN ON DECK AND WELL-DECK LAUNCH AND RECOVERY OPERATIONS BY APPLYING OF QUIESCENT PERIOD PREDICTION (QPP)

B Ferrier, Hoffman Engineering, USA
R J Taylor, Royal Navy, UK
M R Belmont, J Christmas, University of Exeter, UK
L Fedrick, MOD, UK

SUMMARY

One of the most challenging tasks in the deployment of ship-based manned and unmanned VTOL aviation is the launch and recovery and deck handling operational performance on smaller ship platforms. The same may be said of ship Well-deck-based surface boats, like Rhib and LCACs. Awareness and prediction of the ship's position and movement in the ocean is key to ship-based air operations by targeting the right time to deploy and recover. The ships above waterline design coupled by the propulsion systems, affects ship's motion and ship motion induced air wake behaviour.

MoD/University of Exeter's Quiescent Period Prediction (QPP) Program Aims to deliver technologies which support launch and recovery from Naval assets, particularly in higher sea states. The Quiescent Period Prediction (QPP) system achieves this by using a wave sensor system to measure the sea surface several hundred meters in advance of the ship. From the measured sea surface, a short-term deterministic wave model can be constructed allowing the wave system to be propagated to the ship's location. The ship's response to the wave spectrum is calculated to determine the level of quiescence at the time of encounter. The results are mapped dynamically on speed-polar visual graphics providing the ship's best course to steer to ensure encountering the quiescent zone in a timely manner. The ability to fully define a ship's motion as a function of a recovery device, greatly enhances the overall system operational capability whilst fundamentally reducing the inherent risks associated with platform manipulation in high sea conditions. The use of simulation, routinely considered in the UK, as an alternative to traditional approaches such as scale model testing and full-scale sea trials, to predict the safety and performance envelopes of platforms and systems, is discussed. Operating over deck whilst the ship is experiencing a quiescent period normally has the secondary effect of reduced air wake confusion. This is owing to fewer ship structure excursions into and out of the air flow. To better define off deck airflow around the ship a future QPP improvement will be the development and installation of a Doppler LiDAR federate.

The objective of this work is to define ship-helicopter operational deck limits (SHOL), and then to safely expand ship operating deck limits. This article discusses the background to aircraft/ship operations in relation to QPP and goes on to consider both short term implementations (the Landing Period Designator System) and longer horizon deterministically predictive forms of QPP. This article describes the development of the QPP System concentrating on test procedures, timely seaway mapping and ship motion characteristics to complete specific motion sensitive task. An overview of the theory and application is provided concerning dynamic interface analysis with the focus on encountered forces and vehicle to platform response stability.

NOMENCLATURE

DI	Dynamic Interface Study
DOF	Degree-of-freedom
EI	Energy Index (calculated by LPD)
HLA	High Level Architecture
LPD	Landing Period Designator
NATOPS	Naval Air Training and Operating Procedures Standardization
RAO	Response Amplitude Operator
RN	Royal Navy (UK)
RNAS	Royal Navy Air Station (UK)
SHOL	Ship-Helicopter Operational Limit
SMP	Ship Motion Program
SMS	Ship Motion Simulation (NAV routines)
TCS	Tactical Control Station
UAV/S	Uncrewed Aerial Vehicle/System
VTUAV	Vertical takeoff and Landing Unmanned Air Vehicle
WOD	Wind-over-deck

1. INTRODUCTION

The Royal Navy and indeed Navies around the world are changing how they operate aircraft. Uncrewed Air Systems (UAS) are seen by many to be force multipliers and the capabilities they present, once integrated into a digital battlespace will, in time, change the face of maritime warfare. There is an assumption that many of these systems will be attritable, but with budgetary pressures from all quarters and the cost of capability across the military rising can we really afford to waste these systems?

The area of Ship Air Integration is a familiar domain for crewed aircraft but without the pilot in the loop for almost instantaneous decisions and workload assessments the operation and integration of UAS on board is in its infancy. One area the RN is investing in is the development of Quiescent Period Prediction, an application that has utility across all Uncrewed Systems and one that can not only improve safety but increase capability. Awareness of the ship's position and movement in the ocean is key to air operations but to be able to predict this and target the right time to deploy and recover will give Command greater utility and flexibility, something every Commanding Officer would wish for. This operational challenge is owing to the random nature of seaway created deck motion coupled with ship structure disturbed air wake patterns.

MoD/University of Exeter's Quiescent Period Prediction (QPP) purpose is to provide ship motion conditions and motion predictions with sufficient forecasted time (over a minute) to launch, recover and complete other motion sensitive tasks regardless of the seaway. The specific objectives are: (i) to both give the operator guidance on the optimum ship's heading and speed to initiate launch/recovery vehicular activity and (ii) to predict the quietest period of flight deck motion (for safest recovery) over timescales of typically 60 to 120 seconds ahead. to expand ship UAS operating deck limits to approximately Sea State +6.

QPP operates within a federated architecture system allowing plug and play of various measurement instruments with corresponding software wrappers such as a Wave Radar system which is used to map remote sea surfaces several hundred meters in advance of the ship to enable the computation of predictions of the future wave forces acting on the vessel. Such technologies have "look ahead" time horizons measured in minutes and are typically designed to identify "Quiescent Periods". These are "the safest recovery periods" for launch and recovery and make it possible to safely execute an operation when the "mean" prevailing conditions exceed the traditionally considered safe limits.

From the measured sea surface, a short-term deterministic wave model can be constructed allowing the wave system to be propagated to the ship's location. The ship's response to the wave spectrum is calculated to determine the level of quiescence at the time of encounter. The results are mapped dynamically on speed-polar visual graphics providing the ship's best course to steer to ensure encountering the quiescent zone in a timely manner. This procedure defines the ship motion aspect of the deck environment.

This article describes the MoD/Royal Navy sponsored development of the UK's QPP System with discussions on test procedures, timely seaway mapping and ship motion characteristics specifically to complete a motion sensitive task like UAS launch and recovery. Focused discussions on the "proof-of-concept" at-sea trials displaying test results recorded from a standard wave radar, are presented. Data handling and conversions used to create the "best course to steer" dynamic diagram is shown.

Operating over deck whilst the ship is experiencing a quiescent period vessel motion normally has the secondary effect of minimized air wake confusion. This is owing to fewer ship structure excursions into and out of the air flow.

2. QUIESCENT PERIOD PREDICTION (QPP)

A key user requirement of any naval aviation or small surface embarked boat is the ability to launch and recover the vehicle in high sea states. In high sea states it is known that there are periods of time when a ship's motion alternates between large and more modest excursions. One of the means to improve the safety of high sea state recoveries is to predict when such quiescent periods of ship motion will occur and their duration.

Quiescent Period Prediction (QPP) is a class of technologies that use remote sea surface profile measurements to predict the future wave forces acting upon the vessel. Such technologies have prediction time horizons measured in minutes and offer predictions of the actual motions of the vessel. This type of technology is designed primarily to identify so called "Quiescent Periods" in the next few minutes, which make it possible to safely execute an operation when the mean prevailing conditions exceed the traditionally considered safe limits.

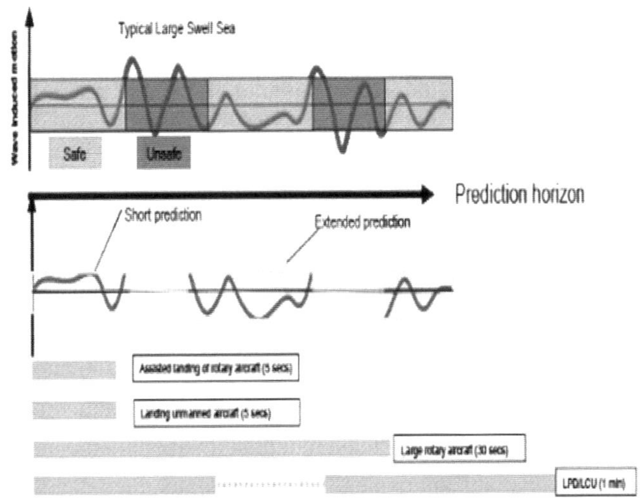

Figure 2.1 – Relationship between Components of Wave induced motion and prediction

A QPP system achieves this by using a wave sensor system to measure the sea surface several hundred meters in advance of the ship **[1]**. From the measured sea surface, a short-term deterministic wave model is constructed allowing the wave system to be propagated to the ship's location **[2]**. The ship's response to the wave spectrum is calculated to determine the level of quiescence at the time of encounter (figure 2.1).

This is a continuous process and can be self-monitored. The components that make up a QPP system are shown in figure 2.2.

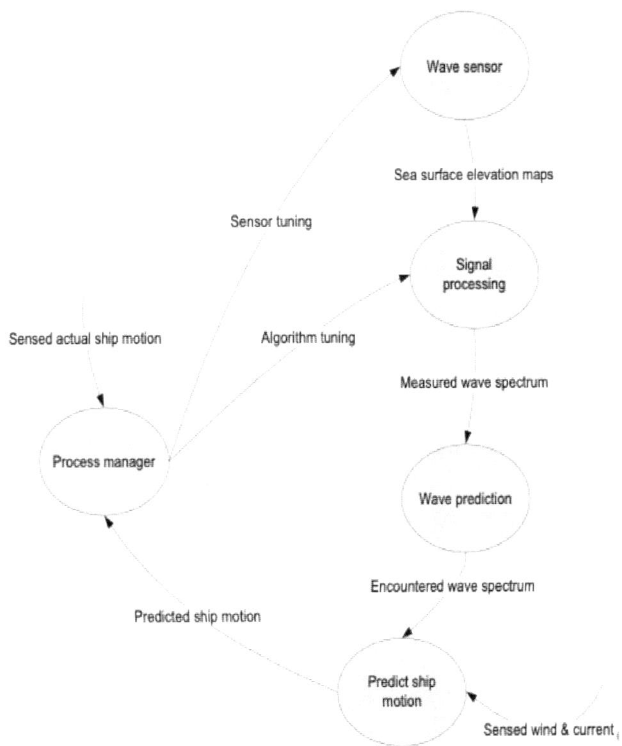

Figure 2.2 – Components of a QPP system

The QPP purpose is to provide ship motion conditions and motion predictions with sufficient time to launch, recover and complete other motion sensitive tasks regardless of the seaway. The objective is to expand ship operating deck limits to approximately Sea State +6. Based on lessons learned from the use of the QPP, it is conceivable, that once integrated, shipboard communities could operate in much higher sea conditions recovering the air/sea vehicles with elevated confidence and reliability. The basis of a QPP system is to use a wave sensor to measure the wave field ahead of the ship, so that ship's motion can be predicted well ahead of the wave field reaching the ship. The system is designed to expand ship operating and deck recovery envelopes. The computational flow diagram corresponding to QPP system components is displayed in figure 2.3.

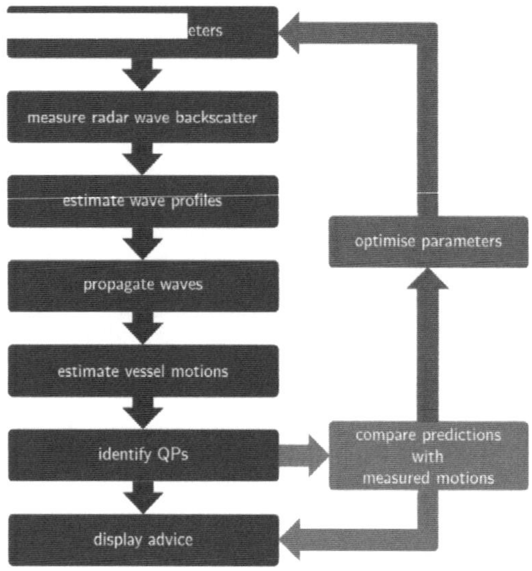

Figure 2.3 – QPP-type Computational Flow Diagram

A key part of the system is the wave sensor device, which can be composed of Light Detection and Ranging (LIDAR) or Radio Detection and Ranging (RADAR). A LIDAR system will require a mechanism to scan a laser beam over the sea surface, whereas a radar system can cover a larger area with a single radar pulse. The performance of a candidate sensor will need to be evaluated in a trial, and if proven, the data collected can then be used to validate a simulation of a complete QPP system.

Each of these systems and related air vehicle operational components are federates, along with the measured environment, which is represented as federates, too. In an initial stage, the measured federates would be simulated. A typical HLA design defines 6 plug and play federates (Figure 2.4), connected via the HLA Run-time Infrastructure (RTI) software. The structure is applicable to either a crewed or uncrewed scenario.

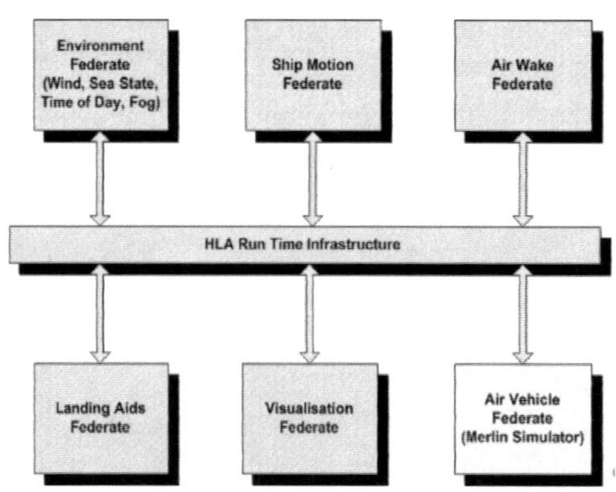

Figure 2.4 – Federation Architecture

The QPP software package is designed to be integrated into a full motion simulator (in the crewed case) and a TCS (in the UAV case). This is a means to estimate system effectiveness as a function of simulated ship motion, visual environment, and synthetic operational systems, and to compare the results to related recorded at-sea data **[3]**.

Quiescent Period Prediction (QPP) systems require: a remote wave measuring system, typically a wave profiling RADAR, a powerful intelligent decision-making computer system and advice aids for human operators or contained within the flight control system of unmanned systems. They typically also include a Landing Period Designator (LPD) sub-system to provide an emergency last minute wave-off capability.

2.1 QPP Process

It is important to recognise the difference between simulation time and processing time. Simulation time can run at any convenient clock speed, real time if human in the loop included faster than real time if Monte Carlo type simulations are required. However, the processing time must be accounted for correctly. The key issue here is to note that any computational time taken in these QPP processes will erode the prediction horizon. For example, a wave of length 150m is capture by the remote sensor around 1 km away. The so-called phase velocity of that 150 m wave is approx. 14.8 m/s, so takes 75 s to travel the 1 km. Thus, any time taken in processing the measured wave and turning it into a prediction vessel motion will erode that 75s. This erosion of prediction horizon must be accounted for in the simulation.

2.1.1 SIGNAL PROCESSING

One of the aims of this aspect of the simulation is to emulate the process by which the output from the wave sensors is analysed to produce a 2D wave surface at the measurement location.

The spectral sea state parameters derived by WAMOS II shown in Figure 2.5 (a) describes the stochastic conditions of the local ocean. These parameters do not contain information about the spatial behaviour of individual ocean surface waves. The process by which the surface elevation map shown in Figure 2.5 (b) is created from the RADAR backscatter image by, effectively, using the inverse of the process used in simulation to generate the RADAR image from the synthesized wave surface.

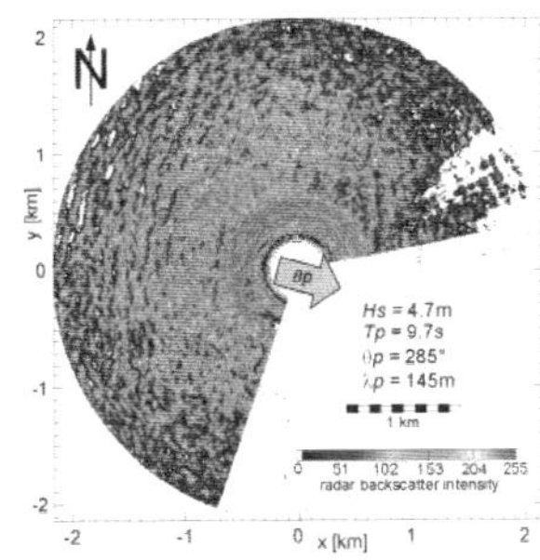

Figure 2.5a: WAMOS II Radar Image

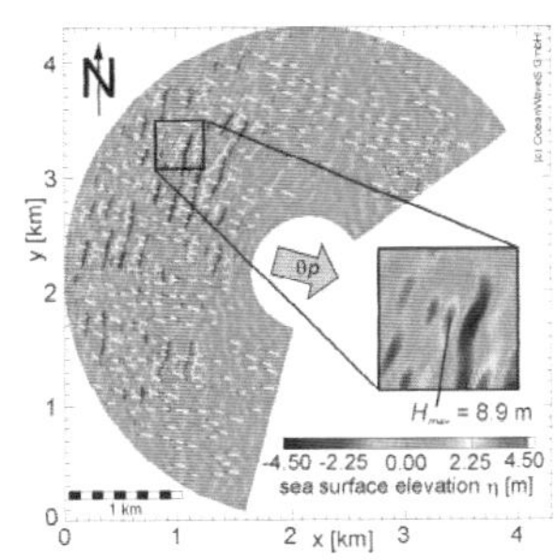

Figure 2.5b: Surface Elevation Image

A second role of signal processing arises because computational costs force the wave predictor model to be strictly periodic (the sea is almost periodic, not strictly periodic). Thus, sections of the spatial wave data must be selected that are most closely periodic, in order to reduced so called leakage errors.

A third aspect of signal processing is necessary if Shallow Angle Wave LIDAR rather than radar is employed because such LIDAR wave data is inevitable non-uniformly distributed in space. Thus, the wave data must be re-mapped onto a uniformly distributed equivalent.

2.1.2 WAVE PREDICTOR

This part of the simulation takes the wave components measured, by the sensors, at some remote location, and propagates those components to the point of encounter by the ship. The ship is moving with forward speed and so the wave surface is varying both temporally and spatially. The theoretical model used to describe the propagation of the wave field is based on classical wave theory.

2.1.3 VESSEL PREDICTOR

Once the wave elevations are known at the point of encounter, in order to determine quiescent periods in the motion, these wave elevations need to be converted to the motion responses of the ship that are pertinent to a particular operational scenario. In the simplest case these are the 6 DOF rigid body motions (surge, sway, heave, roll, pitch, and yaw). Again, since this aspect is part of the QPP process, it needs to be as efficient as possible and come with minimal computational overhead. The reason why the vessel predictor is required to repeat the calculation over the whole of the prediction horizon is that it is expected that the prediction errors would reduce as the prediction time approaches zero **[4]**. Therefore, a computationally efficient way of predicting the vessel motions is to consider linear transfer functions or Response Amplitude Operators (RAOs). These linear responses can be generated beforehand using classical ship motion calculation techniques for use in the simulation.

2.1.4 PROCESS MANAGER

The Process Manager is a distributed artificial intelligence facility that is central to the QPP system; it is not controlling the simulation *per se*, which is more appropriately called the Simulation Manager, but controls real time decisions on parameter values used in the various sub systems and also the flow of information between the QPP processes. It is a pivotal part of a real QPP system not just necessary as part of the simulation. In many cases the QPP process will use parameters as inputs to control the process itself. By monitoring the quality of the predicted motion with that measured by the motion sensors, these input parameters can be optimised for a certain shipboard operation and wave conditions.

2.2 APPLICATION PROGRAM OBJECTIVE

The long horizon forecast program objective is to produce a design for a sea going QPP system based around an intelligent decision-making artificial intelligence wrapper (the AI shell) that exploits data from existing proven hardware and can learn from its previous activity. It applies verified short horizon prediction LPD indices in mapping operationally acceptable seaway regions. Short horizon prediction LPD is described later in the paper.

The QPP system delivers three main benefits:

i. It allows a speed-polar representation of the seaway around the vessel permitting ship's crew to modify ship's heading and speed to ensure the best ship's conditions to complete motion sensitive tasks at hand.
ii. It identifies the safest, most quiescent, period in the vessels wave driven motion over the next minute which corresponds to the safest instant to execute launch/recovery.
iii. The QPP system will incorporate an operator display of the existing confidence level in the predictions. This will allow the operator to appreciate when the system is at its most reliable and avoids the problem of "it works sometimes but we are not sure when".

The QPP system is composed of an AI Shell and LPD measured motion information that accepts data from existing deck interface measurement devices, validated in sea trials and or deployment, thus reducing the need for extensive certification and testing formalities.

In almost all fully integrated technology transition cases, the integration to various independent components presents a high-risk factor. The QPP phase 1 work used existing data recordings, representative ship transfer functions along with existing measurement modular devices. The technical risk will be mitigated through measured simulation and verification lab-based operation Qualification and Systems Integration Tests as well as a shore based QPP device test build-up before proceeding to underway qualification. Central in QPP instrumental components involves the integration and application of wave measured radar system.

3. AIR VEHICLE ON DECK DEFINITION

Motion of an aircraft on the flight deck is calculated in terms of ship motion as a function of the aircraft model. The aircraft model is considered an extension of the ship. The model is defined by aircraft landing gear footprint; on-deck location and orientation; aircraft weight, inertias, center of gravity, lateral drag area and center of pressure. The aircraft experiences ship transferred forces and moments that create rectilinear and angular accelerations on the air vehicle. The accelerations can be numerically integrated to determine the position and attitude of the helicopter relative to the ship as function of time, for various ship motions **[5]**. In essence, the aircraft is displaced as the sum of all forces to which it is exposed. Figure 2.6 displays a resumé of the moments acting on a skid-based aircraft at sea.

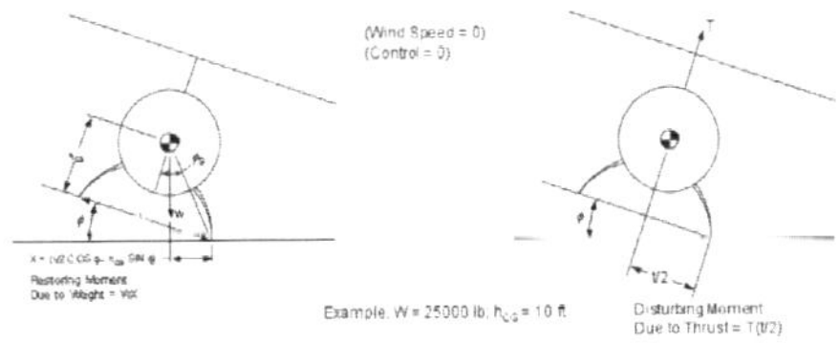

Figure 2.6: Moments Acting on Skidded Helicopter Sea

The inertial loads at the helicopter center of gravity induced by ship motion are given by.

$$
\begin{aligned}
F_{i_x} &= W * AXcg \\
F_{i_y} &= W * AYcg \qquad (1) \\
F_{i_z} &= W * AZcg
\end{aligned}
$$

where, $\begin{pmatrix} F_{ix} \\ F_{iy} \\ F_{iz} \end{pmatrix}$ = inertial forces due to ship motion

W = aircraft weight

$\begin{pmatrix} A_{xcg} \\ A_{ycg} \\ A_{zcg} \end{pmatrix}$ = acceleration

In the longitudinal, lateral, and vertical directions, these inertial loads become:

$$
\begin{pmatrix} X \\ Y \\ Z \end{pmatrix} = \begin{pmatrix} T_{11} & T_{12} & T_{13} \\ T_{21} & T_{22} & T_{23} \\ T_{31} & T_{32} & T_{33} \end{pmatrix} \begin{pmatrix} F_{ix} \\ F_{iy} \\ F_{iz} + W \end{pmatrix} \qquad (2)
$$

where: Tij = T (ϕ, θ, ψ) (transformation matrix from ship's axis system to horizontal level/vertical axis system).
And,

ϕ = roll
θ = pitch
ψ = yaw

A wind force is added to the ship motion induced forces. In SMS, a unidirectional continuous wind model (simplistic model), with its vector in the same direction as the seaway, is applied. Figure 2.7 displays the inertial components associated with the introduction of wind on the computation of the sum of forces at the center of gravity of the air vehicle. The wind vector is defined by its magnitude (Vwod) and its direction (Dfwod). To compute the lateral force applied at the aircraft Center of Pressure due to the wind, the Vwod is resolved along the normal to the aircraft center line ($V_{W_{long}}$ and $V_{W_{lat}}$). The lateral component is used to compute the lateral force, as follows:

$$F_{W_y} = 3.5\,Ay\left(\frac{V_{W_{lat}}}{100}\right)^2 \qquad (3)$$

where:

Ay=Aircraft projected area normal to the $V_{W_{lat}}$ component

F_{W_y} =Lateral force applied to aircraft center of pressure due to wind

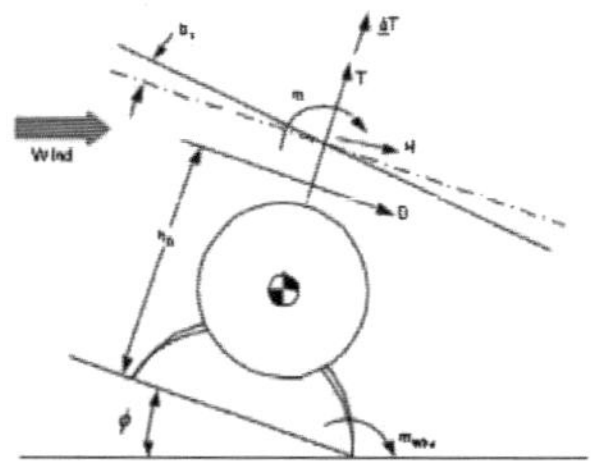

Figure 2.7: Wind Acting on Skid-based Vehicle on Deck

The standard tricycle coordinate system used to develop the stability of air vehicle on the deck of a moving vessel is displayed in Figure 2.8. In this study, the helicopter handling system probe replaces the nose wheel, and the skids are reduced to point references replacing the main gear.

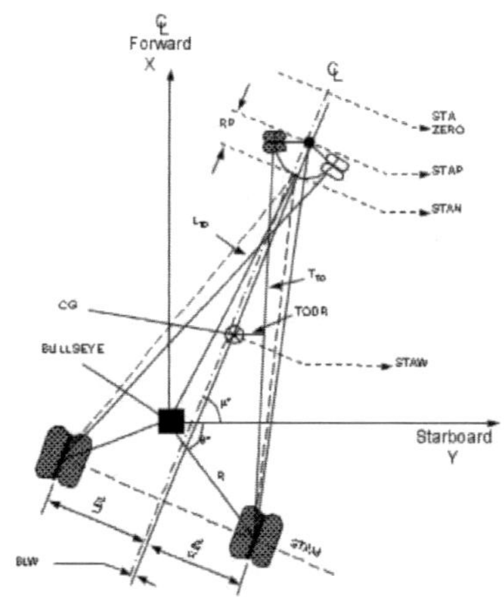

Figure 2.8: Typical Air Vehicle Freebody Diagram

The axial forces on the main landing gear due to the wind force F_{W_y} is given by:

$$F_{RMG_{WIND}} = (F_{W_y})\frac{(WL_{CP} - WL_G)}{(L_{BL} - R_{BL})} \qquad (4)$$

where:

$F_{RMG_{WIND}}$	= Main Gear (right) axial force
F_{W_y}	= Wind lateral force component
WL_{CP}	= Center of pressure waterline
WL_G	= Ground waterline
L_{BL}	= Left wheel butteline
R_{BL}	= Right wheel butteline

The incremental aircraft roll due to the wind is given by:

$$Df_{(WIND)} = \tan^{-1}\left(\frac{F_{RMG_{WIND}}}{(K)(L_{BL})}\right) \qquad (5)$$

where: K = spring constant

Axial forces on the main landing gear due to aircraft inertial forces in the plane of the main gear, is given by:

$$F_{RMG(Inertia)} = Y\left(\frac{WL_{CG} - WL_{G}}{L_{BL} + R_{BL}}\right) \qquad (6)$$

where:

WL_G =Center of gravity waterline

$F_{RMG(Inertia)}$ =Right main gear axial force due to the lateral inertial force Y defined in equation (5).

Assuming perfect rocking, the axial force on the left main gear is opposite to the force vector acting on the right main gear:

$$F_{LMG(Inertia)} = F_{RMG(Inertia)} \qquad (7)$$

where: $F_{LMG(Inertia)}$ = Left main gear axial force

The incremental aircraft roll due to inertial loads is determined by:

$$Df_{(Inertia)} = \tan^{-1}\left(\frac{F_{RMG(Inertia)}}{(K)(L_{BL})}\right) \qquad (8)$$

The simulation model assumes a constant wind, therefore, $Df_{(wind)}$ is constant throughout the simulation run. However, $Df_{(Inertia)}$ is continuously changing with ship motion. The total incremental change in the aircraft roll with respect to the ship is given by:

$$Df_{(total)} = Df_{(wind)} + Df_{(Inertia)} \qquad (9)$$

Turnover incidents are static or dynamic in character. Static turnover is the same as on shore. The resolved weight vector migrates beyond either the friction forces causing the aircraft to displace or the reaction forces causing the aircraft to turnover. Figure 2.9 develops dynamic turnover caused by the rotor disk (uneven loading of the rotor) or by ship motion, the same phenomena occur.

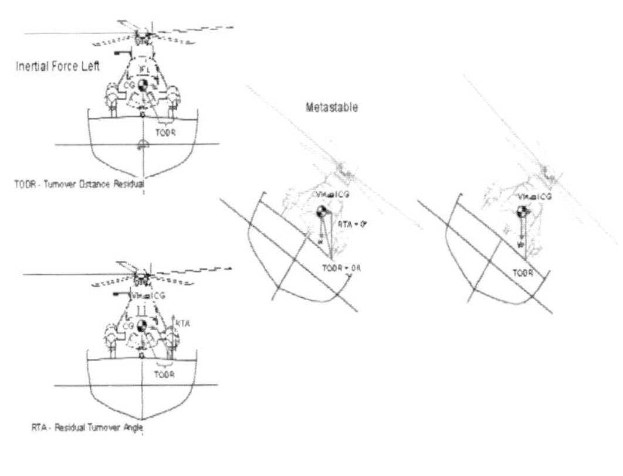

Figure 2.9: On-deck Dynamic Turnover

The aircraft center of gravity is in motion. In the sum of forces, the weight vector is continually modified in response to inertial forces applied by either the rotor disk or ship motion or both. The distances TODR, TODL, and PBD essentially reflect system stability. At the point where a distance becomes negative, the system is unstable and will seek to find a more stable, but usually undesirable geometric solution. In similar fashion, when the landing gear friction values are exceeded by the combination of aircraft apparent weight and induced inertial forces, slippage will occur. Aircraft slide will continue until the aircraft frictional forces are greater than the disturbing inertial forces. Finally, when the vertical inertial force equals and opposes the aircraft weight, the deck friction goes to zero and an unintentional liftoff is indicated. The sum of these incidents trace aircraft-ship operational envelopes.

2.1 VEHICLE – DECK INTERFACE

For the helicopter to operate to the deck without a deck officer, it needs to know what the ship is doing now and what it will be doing during the final descent to touchdown. The Landing Period Designator (LPD) Energy Index (EI) algorithm is designed to identify quiescent periods of ship motion suitable for the recovery of the helicopter (figure 2.10).

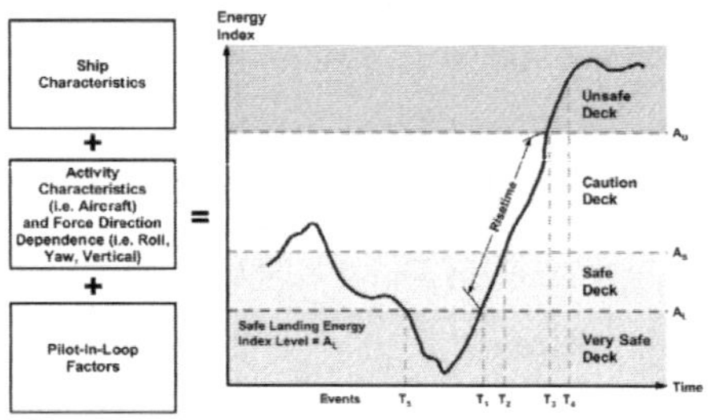

Figure 2.10: Deck Status and Rise time

The time required for the deck motion to rise from minimal motion (or very safe deck) to unacceptable motion is called the rise time. In terms of the EI scale, the rise time is defined as the period that is measured from the end of a green signal to the positive side of the red line. This is given as ($T_3 - T_1$) as shown on Figure 2.10. The rise time is a thumb print characteristic of the ship's response and remains constant for each ship class.

The very safe deck is a special condition in which there is insufficient energy in the aircraft-ship system to raise the deck out of limit for some defined time period or rise time, and it is this concept that was used in the simulators to indicate that the deck was safe to initiate landing.

Assuming the minimum average rise time for a given vessel remains constant, the aircraft descends within the rise time value and the aircraft is assured a recovery within the deck motion limits. Essentially, the methodology summarized above is a formulation to quantify operating beyond the static deck limits as defined in NATOPS or SHOL. Figure 2.11 displays graphically beyond static limit operations. The base envelope is taken from 10 knots of ship speed. As before, any points within the hour-glass structure are conditions within limits and contain no appreciable probability of out-of-limit deck motions. Outside of the structure contains motions which are considered by static reference as out-of-limit **[6]**.

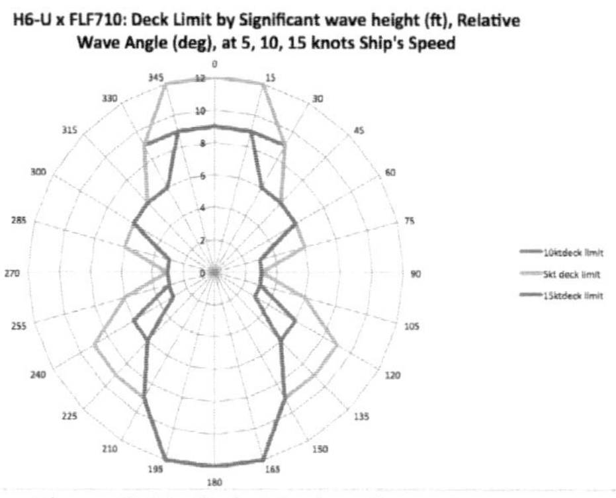

Figure 2.11: Safe Motion Operations

This is particularly important for vehicles without pilots seated in them (Unmanned Air Vehicles or UAV). A schematic representation of how the LPD is used for UAV autorecovery to signal the onset of the descent (and of deck quiescence) is displayed in Figure 2.12. Throughout the approach and initial hover, LPD is monitoring the deck. It is only at the low hover that the UAV autoland system would accept a Green Deck signal from the LPD. When the EI signals the onset of quiescence and through the uplink sets the air vehicle on its descent to the deck. Figure 2.12 displays current VTUAV Flight Patterns.

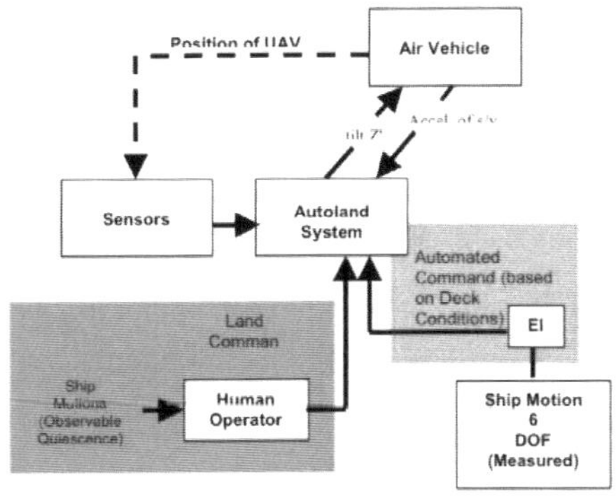

Figure 2.12: Notional Autoland System

2.2 IMPACTS OF DECK MOTION ON DECK LIMITS

The simulation of ship platforms and systems to assess their interfaces to other platforms and systems need to be made more available. Early use of trusted simulation can allow major ship projects to be de-risked during design to a greater degree than has been possible in the past. During ship design phases, the performance of platform operations and platform interactions such as helicopter recovery, replenishment at sea, vehicle docking, and vehicle deployment and recovery can now be assessed in all required sea and weather conditions **[13]**. Establishing the accuracy of simulations developed for supporting design is vitally important perhaps more so than for simulations that have been developed specifically for training. Numerical inaccuracies in training systems can normally be identified early by experienced operators (not necessarily my experience, particularly with respect to ship air wake).

The calculated system limitations provide experimental DI with the necessary data to more effectively set testing strategy to probe the limiting conditions **[7]**. Continuing the theme of ship's motion and having created a large library of data on the target ship, it is possible to derive tendencies of ship's characteristics to the most basic of parameters, such as, ship's speed. Figure 2.13 displays the effect that ship's speed has on roll motion (here assessed at 10 feet significant wave height).

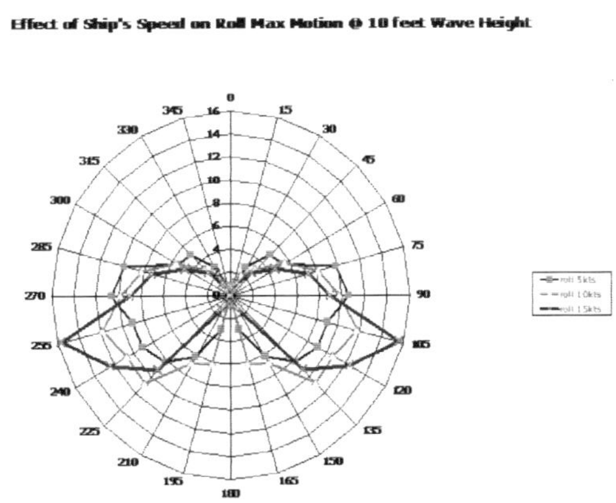

Figure 2.13: Effect Ship's speed on Roll Motion

Roll appears to improve slightly with increasing ship's speed in ahead seas through 90° relative wave angles. From midship to 135° relative wave angles, roll angle becomes larger with increasing ship's speed. It then falls away at angles greater than 150° to 180° relative wave angles. Studying pitch (see figure 2.14), the tendencies are opposite those of roll.

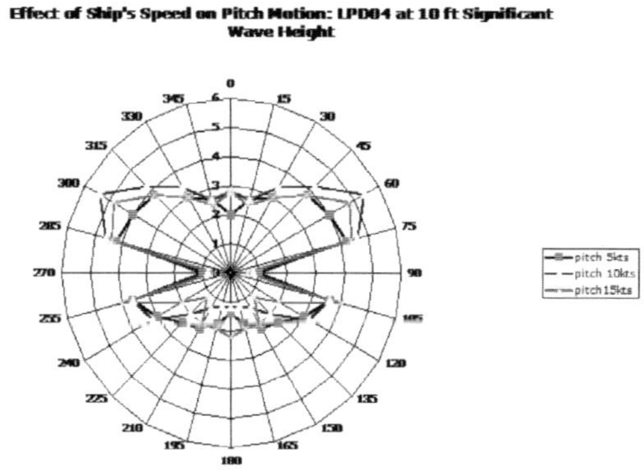

Figure 2.14: Effects of Ship's Speed on Pitch Motion

Much as expected, in the ahead seas condition pitch motion increases with increasing ship's speed. The following seas cases are more complicated, but overall follow the same theme.

Assessment was made concerning landing gear responses to encountered deck loads from several ship models. Identification of incidents of gear slide (coefficient of friction = 0) and gear dampener stroke are produced. This is done for each motion scenario. A sample analysis is given in Figure 2.15. The example was randomly taken from data file 10301511, which translated means, 10 knots ship's velocity, and 30 degrees relative wave angle 15 feet significant wave height and 11 seconds wave period. Calculated encountered axial forces are shown over the course of the time history. Fz (vertical forces) is the largest contributor to the resultant force vector (Fr). For this scenario, Fx (longitudinal force vector) is statistically slightly greater than those of Fy (lateral force vector). At the point where the coefficient of friction measures zero (or the index of slide is greater than one), an aircraft displacement is indicated. The sum of the force distribution studies produces a helicopter encountered force deck limit diagram (rondelle, speed polar or O'Reilly plot).

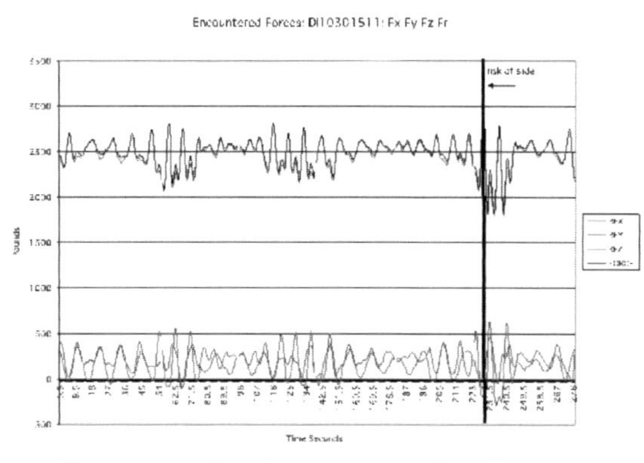

Figure 2.15: Landing Gear and Probe Encountered Deck Forces, File 10301511

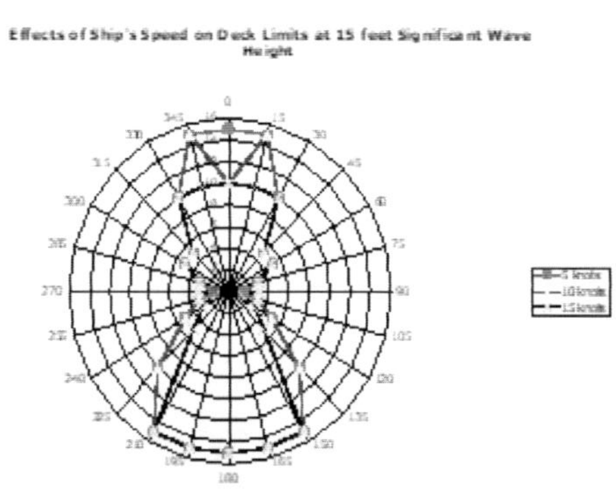

Figure 2.16: Encountered Deck Forces

As developed earlier for vehicle interface simulation, Figure 2.16 represents deck limits as a function of deck encountered forces, in this case onboard the LPD-4 running at 5 to 15 knots of ship's speed. Reading the diagram is like that for aircraft stability. Anything inside the hourglass figure is within limits where there no measurable risk of aircraft incident recorded. Anywhere outside the hourglass, an incident of landing gear faults was recorded and is, therefore, out of limit. As an example, at 60 degrees relative wave angle, seaway significant wave heights lower than 5 feet are within limits. In interpreting the diagram, it should be apparent that the large size of the Austin Class ship has a great deal to do with deck stability. Ahead seas show good stability between 330 – 360/0 – 30 degrees relative wave angles and in following seas 150 – 180 – 210 degrees of relative wave angles. Helicopter displacement risk appears to be focused on the beams. Areas without risk on the beams occur with less than 3 feet significant wave height (or sea state three). The most heavily constrained areas are formed from 60 – 120 degrees of relative wave angles and from 240 – 300 degrees of relative wave angles.

The effects of ship's speed on the deck limits are displayed in Figure 2.17. The interface between the LPD-4 response characteristics combined with the Fire Scout aircraft stability parameters, appears to be particularly sensitive, with minor adjustments in following seas, to ahead seas at higher ship's speeds. From the diagram, ahead seas (330 – 30 degrees relative wave angle) is most stable at 9 feet significant wave height. The stable condition quickly decays at relative wave angles greater than 030 – 135 and 225 – 330 degrees of relative wave angles.

2.3 IMPACTS OF DECK MOTION ON VEHICLE DECK LIMITS

Figure 16 is composed of the launch and recovery events, energy index, ship's roll, pitch, and yaw traces along with the deck energy levels. From the time history, with LPD off, a normal evolution had taken place with both the launch and recovery events occurring on a green-amber or safe deck, but not quiescent or green deck. The other channels analysed, corresponding translational traces, engine torque and landing gear compression all appeared to have shown normal operation.

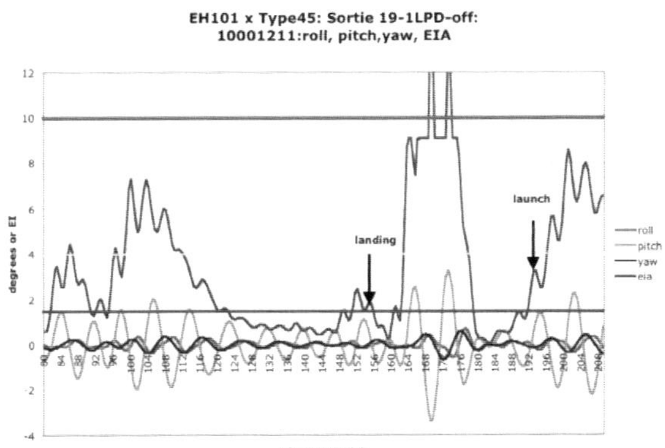

Figure 2.17: Trial (Day) with Launch and Recovery

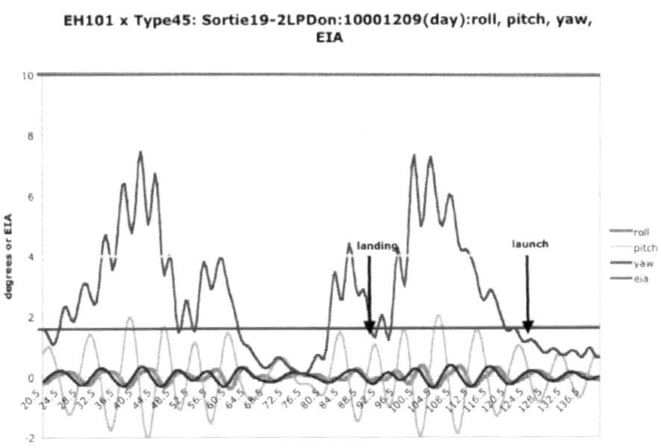

Figure 2.18: Trial Day with Launch and Recovery

From the time history in Figure 2.18, with LPD on, a normal evolution had taken place with both the launch and recovery events occurring on a green or quiescent deck.

The other channels analysed, corresponding translational traces, and engine torque appeared to have shown normal operation. Oleo compression trace appeared to show a peak compression on the nose gear (Figure 2.19).

Nose wheel

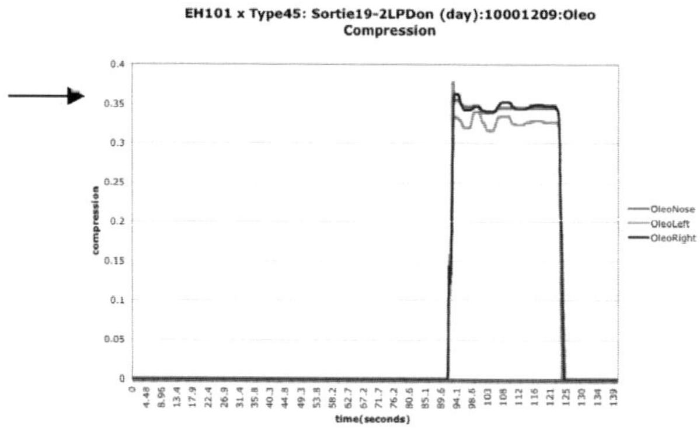

Figure 2.19: Sortie Oleo Compression Trace

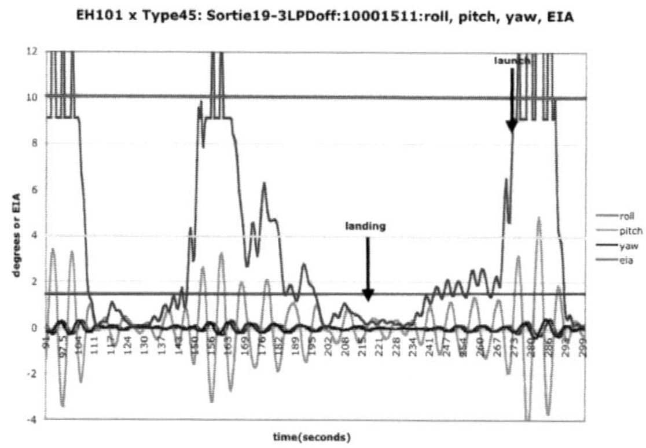

Figure 2.20 – Trial Day with Launch and Recovery

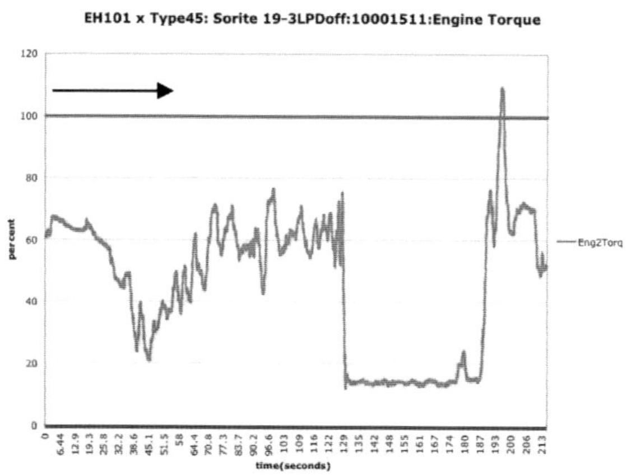

Figure 2.21: Sortie 19-3 Engine Torque

Figure 2.20-time history, with LPD off, the recovery event occurred on a quiescent deck, but the launch happened from a high amber or caution deck. The corresponding translational traces showed similar large displacements at launch. Oleo compression appeared normal, but the engine torque was measured greater than 100% (figure 2.21).

3.3 SHIP – SHIP NATO SUBMARINE RESCUE; *similar physics longer launch and recovery time*

The NATO Submarine Rescue System (NSRS) managed by the UK MOD, Defence Equipment & Support. In the event of a serious submarine incident, a Submarine Rescue Vessel (SRV) would be dispatched to rescue and transport submariners from the distressed submarine to a surface ship also referred to as the Mothership (figure 2.22).

Figure 2.22 – NSRS UK Version

A key user requirement of this system is the ability to launch and recover the SRV in up to 5 metres significant wave high (SWH). This includes the launch and recovery of a RHIB tender (figure 2.23). However, proving this capability at sea carries a significant risk. A series of validation tests has been undertaken using sea trials data to validate a related simulation to predict scenarios that cannot be viably tested.

Figure 2.23 – NSRS Tender RHIB

The QPP Landing Period Designator (LPD) stand-alone unit (here known as LPD1) to collect accurate MoShip ship motion data, and provide short term MoShip ship motion prediction, in particular, was operating throughout the trial **(16).** An additional Motion Data Logger (similar to the LPD stand-along unit) was used to collect accurate SRV and RHIB motion data. There are no aids on most surface combatant ships to help the operator in anticipating a ship's deck quiescent period, either long or short horizon dimensions. The LPD is designed to provide short horizon ship motion cues to distinguish ship deck quiescent periods that result in acceptable conditions for a shipboard capsule recovery or launch **(17).** The LPD component is designed to fill a QPP technological gap related verified short horizon predictions during all-weather operations.

For most of the QPP/NSRS trials, the objective of using the LPD is to document deck motion characteristics as it may affect the SRV shipboard launch and recovery capabilities. Equally crucial, is the verification that submersible capture activities occur within safe motion parameters (short horizon motion computation). The Energy Index, the operative algorithm of the LPD, identifies quiescent periods to initiate vehicle acquisition based on vehicle deck limit definitions.

MoShip energy index ship motion responses are calculated as a function of the submersible rescue motion criteria. The submersible model is considered an extension of the ship. The vehicle experiences ship transferred forces and moments which create rectilinear and angular accelerations on the vehicle. The accelerations can be numerically integrated to determine the forces and attitude of the captured vehicle relative to the ship as function of time, for various ship motions.

QPP based computations centered on MoShip and RHIB ship interface performance criteria not including the RHIB handling system. Once the automated RHIB crane launch and recovery system limits are known, it is possible to re-evaluate its performance using the recorded MoShip motion data as a post-processor exercise.

Using the information regarding the sensor operation and interfaces collected during this trial, a QPP simulation design specification is possible based on a High-Level Architecture (HLA) operation system.

4. QPP PRE-PROTOTYPE SEA TRIALS

Figure 4.1 shows the main data generation elements in QPP systems which are incorporated into the quiescent period prediction process. The primary inputs arise from the wave sensor system, i.e., RADAR or LIDAR. In contrast the other inputs such as vessel motion, the GPS vessel position designator, wind speed, and data-base values such as NSRS operational criteria, etc. provide parametric inputs used by the process manager for adaptive decision making on various aspects of the prediction model. The exception being the LPD short horizon prediction for which vessel motion is the primary input. In QPP, the LPD functions as a real-time motion sensor ensuring that critical tasks are conducted within specified motion criteria. Two main processes operate on a time-shared basis during QPP. The first is a statistical algorithm which identifies the direction power spectrum of the waves, in order to provide wave directions to the deterministic prediction model. The deterministic prediction process itself uses shorter data sets, of the order of 50 seconds, and has an execution cycle time of at most a few seconds. This allows the prediction cycles to be overlapped providing multiple assessments of the predicted waves at the NSRS site and thus allowing the process manager to reject anomalous results. If the only motions existing are pitch and roll, then on recovery (or launch) the MoShip deck would appear essentially as a sloped static platform. The ship's (or any floating body) displacement is in fact dynamic and moves in 6 degrees-of-freedom, with force driven rates and accelerations. Deck roll and pitch positions of a stabilized platform are achieved by forces caused by deck rates and accelerations which can be measured by degree-of-freedom. Thus, not only are the static angular deck position a concern, but also the corresponding translational rates, which in turn affect recovery ship-based lever-arm deflections.

Motion of a vehicle on the deck is calculated in terms of ship motion as a function of the vehicle model. The SRV model is considered an extension of the ship. The aircraft experiences ship transferred forces and moments that create rectilinear and angular accelerations on the vehicle.

More than 350 motion time history hours were recorded over the three-week trial period. More than 7 hours were recorded on either the SRV or NSRS RHIB. There were 350 hours of recorded motions typical for sea states calm to +6. The test ship encountered conditions more than 6 metre seas along with sustained winds measuring nearly 40 knots. The higher seas occurred during the QPP segment of the trial. Ship's motions were recorded in the NSRS segment between 12 Nov to 22 Nov. SRV launch and recovery events were recorded on the 18th and 20th Nov, both during low sea conditions. Several the SRV training exercises involved Scripps wave buoy launch and recovery, as well. This confirmed the functionality of the buoys and precision of recording the low sea state conditions. Between the NSRS and QPP segments, the NSRS SRV equipment was removed removing a significant weight off the stern of the Northern River (MoShip). QPP motion data were recovered between 23 Nov – 27 Nov. RHIB motion was recorded on 26 Nov. During much of the QPP trials, wave height data were also recorded using the Scripps wave measurement buoys.

As explained earlier, the entire day is evaluated in the full analysis, permitting to display a complete percent distribution of deck stability along with other statistics like max motion by axis. For this preliminary analysis three (3) hours on the same day were selected, one of which contains a RHIB exercise. In the section which follows a file common to both the RHIB and MoShip were chosen [26 November at 10:00]. Figure 4.1 displays the meteorological condition for that time period.

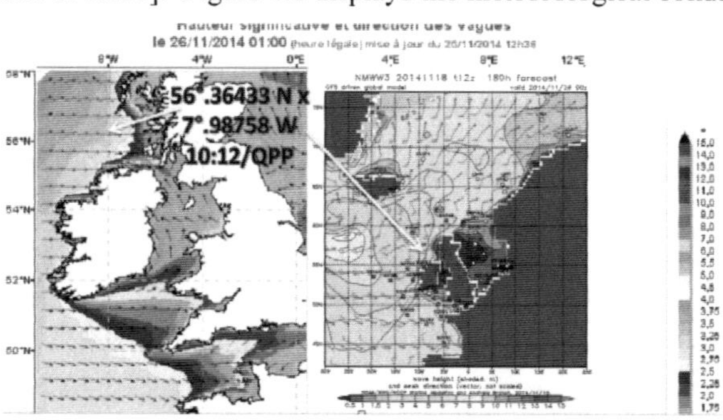

Figure 4.1: Seas in the test zone

Figure 4.1 displays the significant wave heights from both the European and NOAA models for the 26 November. The MoShip travelled during the night from a sheltered area near the Isle of Mull to a zone between the New Hebrides and the Irish border in the North Atlantic. The objective was to expose the QPP trial equipment to the leading edge of an Atlantic storm allowing the test team to pick and choose climatic conditions favourable to the trial order and plan. The ship travelled from 2.5 – 4 metre significant wave heights. This permitted the QPP test equipment to measure well defined waves from a confused to fully developed sea. The seaway choices were elevated but within acceptable operational limits permitting envelope expansion for the launch and recovery of the NSRS trial RHIB. Winds measured between 15 – 25 knots. The ship was not weighed down by the NSRS kit and was riding high in the water. Owing to the expert seamanship of the ship's Command, the ship was almost always on an even keel. As shown in Figure 4.2, the maximum angular displacements recorded were 5.4° roll and 4.6° pitch. The pitch sensitivity may allude to the ship modification which extended the vessel by several metres. The translational rate maximum recorded, Z'= 1.8 m/s, Y'= 0.7 m/s. The corresponding accelerations measured 2.6, 2.1 m/s**2 respectively. The ship motions exceeded deck limits but still permitted NSRS operations (if the SRV was present) for 98.3% of the time history. The SRV deck limits are 5° roll, 2.5° pitch, 7.2 ft/sec (2.2 m/s) Z', 2.8 ft/sec (0.85 m/s) Y'. The RHIB recordings used SRV limits.

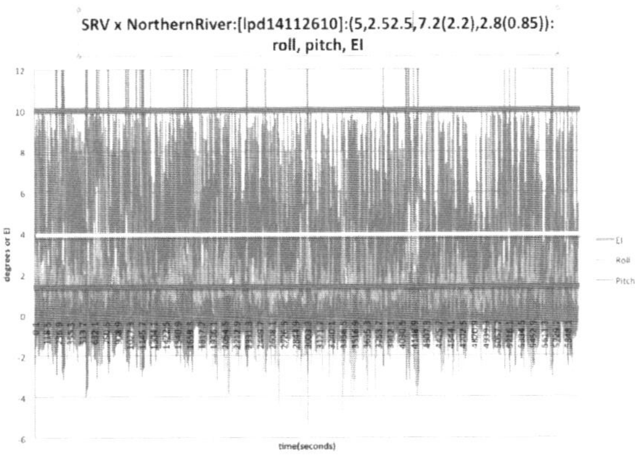

Figure 4.2a: Time History trace recorded 26 Nov at 10:00

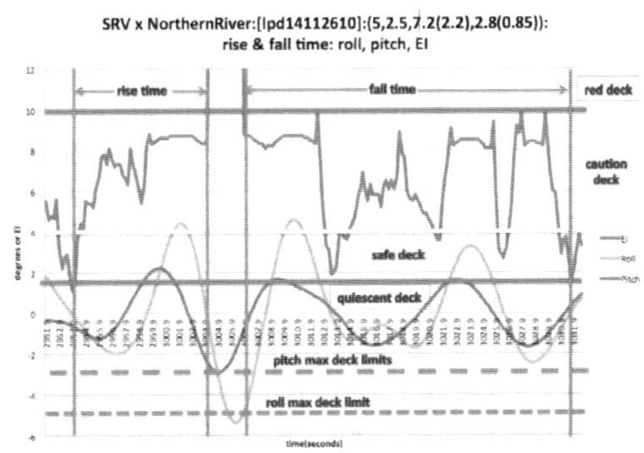

Figure 4.2b: Motion Event @ 29 – 30 mins into run

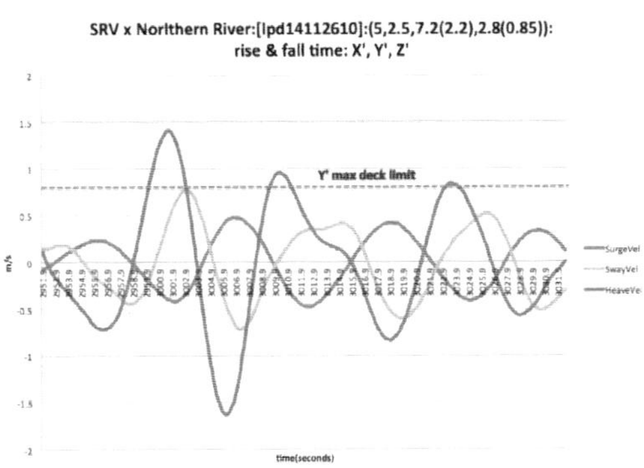

Figure 4.3a: Corresponding rate recorded

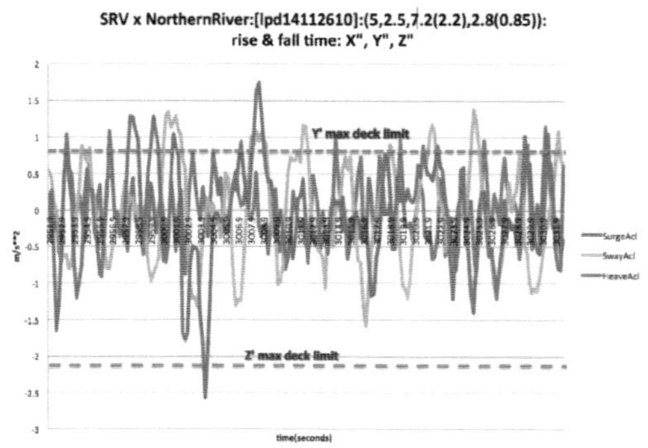

Figure 4.3b: Corresponding accelerations recorded

Figure 4.2 displays a standard one-hour time history data file containing roll and pitch motions along with the corresponding energy index trace. Studying the graphic, it is fairly evident the ship experienced significant motions owing to the rising seaway yet remained well balanced (no departure from trim). Figure 4.2a displays a sample rise and fall time event recorded in the 29 – 30 minute into the time history trace. The rise time (the time it takes for a ship in quiescence to accelerate out of limit) measured 10 seconds in length whilst the related fall time measured 42 seconds. The corresponding lateral and vertical rates and accelerations are shown in figure 43a, b.

Examining the related traces (figures 4.3a and 4.3b), the ship experienced a significant Y" impulses strong to produce excessive Y' motions which, in turn, fueled a significant coupled roll event. Similarly, vertical rates were sufficient to cause excess pitch displacements. The evaluation summary indicates that the deck was available for SRV operations 98.3% of the time. The deck was out-of-limit for over 31 seconds of the recording. The deck was quiescent 6.1% of the time; green, amber (safe) 23.9%, amber (caution deck) at 68.3%. The deck was measured out of limit (red deck) more than 1.7% of the time. There were 24 rise time events recorded all measuring much longer than 6 seconds.

5. QPP BEST COURSE TO STEER

To best describe spatially the measured seas acceptable for flight operations, the energy index is used. It is a short horizon algorithm which is also used as verification that the capture or recovery is occurring within specified limits. As described earlier, one of the computational approaches is to use the QPP algorithms to measure in real-time the seaway around the Mother ship. The converted data are coupled with stochastic ship transfer function information to compute forecasted ship's motions. This is calculated for a given ship's heading and speed. Particular attention is given to the area towards which the ship is heading. It is estimated the data will be computed for a diagram representing the ship's motions up to 2 km from the initial measurement position.

A speed-polar representation of the seaway to guide ship's command and deck hand activity, appears as a synoptic map created by the lay-over of various measured and computed parameters. Each parameter speed-polar lay-over is application specific computed as a function of the acceptability of ship's conditions to complete a specific motion sensitive task (figure 5.1). This is based on NAVSEA SEA 851, formally known as the Seakeeping Department, Active Operator Guidance System's post-processor **[8]**.

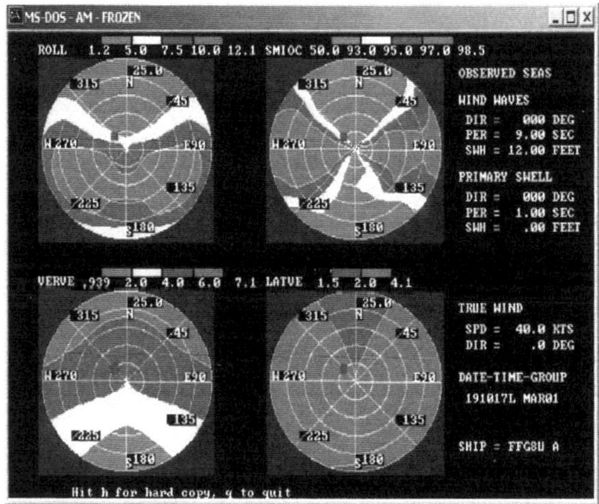

Figure 5.1 – Parameter Map Speed Polar

A limit is defined by the impact that a certain ship motion condition may impose on the air/ship vehicle's structural integrity or that may result in undesired dynamic response. The sum of these lay overs produces an active operator's guidance map (figure 5.2). All the various QPP components either exist or require limited developmental modification.

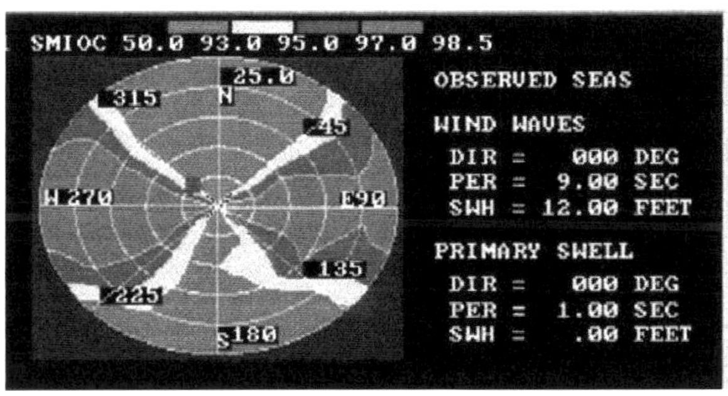

Figure 5.2 – Best Course to Steer

Multiple solution convergence methodologies may be offered using, for example, real-time long horizon forecasting alone or coupled with seaway real-time mapping providing stochastic (ship transfer function) based distance motion forecasts updated several times per second.

QPP can be applied to all ship motion sensitive activities. Throughout air vehicle approach and initial hover, QPP would be producing operator guidance maps allowing the ship to adopt a heading and speed acceptable for air vehicle recovery. It is only at the low hover position that the QPP short-horizon component would be used to verify that the deck was measured in real-time as being in a quiescent period condition.

6. CONCLUSIONS

A significant motivation for the development and deployment of VTOL UAS, is their exploitation in situations or conditions in which human exposure would by highly risky. Awareness and prediction of the ship's position and movement in the ocean is key to air operations by targeting the right time to deploy and recover. MoD and the University of Exeter's Quiescent Period Prediction (QPP) purpose is to provide ship motion conditions and motion predictions with sufficient forecasted time (over a minute) to launch, recover and complete other motion sensitive tasks regardless of the seaway

This paper provides a discussion on the class of technology that use remote sea surface profile measurements to predict the future wave forces acting upon the vessel. Quiescent Period Prediction (QPP) system achieves this by using a combination wave radar sensor and deck motion measurement systems to calculate the sea surface several hundred meters in advance of the ship. The QPP system is composed of an Artificial Intelligent (AI) shell and LPD measured motion information. From the measured sea surface, a short-term deterministic wave model can be constructed allowing the wave system to be propagated to the ship's location.

The paper explains how the QPP radar system is employed. RADAR backscatter images are recorded of the wave system impinging on the vessel are acquired out to typically 2km. The RADAR backscatter data is processed to obtain estimates of the detailed shape of the sea surface elevation. Care is necessary to minimise the effect of artefacts such as the wave shadowing. The sea surface elevation data is used as the input to a linear wave propagation model that estimates the two-dimensional Fourier transform of the future phase resolved directional wave spectrum at vessel location typically up to 120 sec ahead in time. The phase resolved directional wave spectrum is applied to the phase resolved RAO coefficients of various modes of vessel motion. The spectral results are the inverted into the time domain to obtain estimates of the predicted vessel motions. Finally, The predicted vessel motion data is used to inform the vessel operator of both the best course and speed and the time of occurrence and the duration of the forthcoming quiescent periods in the context of the maritime operation of interest.

The future QPP programme will need accurately defined milestones in order to target required technology readiness levels and estimate costs to gain access to limited budgets.

Amongst the targets the programme is expected to include:

- wave propagation modelling.
- radar and lidar wave mapping.
- size of prediction horizon.
- hardware design.
- sea tested breadboard.

The project Quiescent Period Prediction (QPP) aims to extend the prediction of the vessel motion for up to a minute ahead of time. Sea trials have begun to investigate the feasibility of developing such a system, with the core hardware functionality in existence, and the algorithms validated to provide reasonably accurate predictions.

An early overview of the ship motion and dynamic interface have been considered with the emphasis on forces induced resulting on ship stability. Validation of the results is a priority because of the potential problems affecting operating deck limits to be programmed for SRV or the associated RHIB recovery. Beyond the basic problem of data verification and validation, the analytic procedure demonstrated above using the operating components of the future QPP device, is sound and could be used to cross-correlate between proposed SRV-ship deck limits and the vehicle's expected physical responses.

The data precision and computational results appears to be of high quality favouring the validation of the QPP thesis. Complex physics-based simulations have the potential to predict capability such as safe operating limits of equipment in high sea states. The trials results indicate that QPP systems have the potential to further extend the operating envelopes of equipment in high seas where other prevailing met-ocean factors allow. The NSRS simulation currently predicts operating limits for NSRS recovery. Certain QPP equipment has been tested at sea and data logged whilst operating within an NSRS recovery scenario. The project is determining the accuracy and scope of the QPP technology and the extent to which simulation could be used to design and prove a suitable on-board capability.

Early indication is that the current NSRS simulation design can be extended to include a QPP function. This is true even with the relatively coarse predictions from the fully two-dimensional methods used. These computational methods have been shown to be adequate for QPP purposes in the large seas of interest. The resulting simulation would be capable of assessing QPP functionality in terms of performance. Three specific benefits of the simulation would be: (i) to allow the exploration of the effectiveness of QPP over a wide range of sea conditions, (ii) to provide a test platform for upgraded prediction algorithms as they come on stream (iii) as a software design tool for sea going QPP hardware systems.

From the human resources perspective, simulation provides a pathway for gaining understanding of how the QPP information can best be applied by the NSRS operators and the number of other on-board applications potentially benefiting from a QPP system.

In a related but future study activity, it was observed that certain ship deck condition tendencies were apparently related. Operating over deck whilst the ship is experiencing a quiescent ship motion period normally has the secondary effect of minimized air wake confusion. This is owing to fewer ship structure excursions into and out of the air flow. To better define deck airflow around the ship it is proposed to integrate a Doppler LiDAR instrumented federate. The purpose of this is to predict the future vessel air wake and look for quiescent periods in this paralleling the vessel motion QPP technique. Preliminary test results from a proof-of-concept trial, are briefly discussed. The ability to fully define a ship's motion and air wake condition as a function of a recovery device, greatly enhances the overall system operational capability whilst fundamentally reducing the inherent risks associated with platform manipulation in high sea conditions.

7. ACKNOWLEDGEMENTS

The authors wish to pay tribute to the technical competence and outstanding professionalism of retired Dr. John Duncan's former head of SBA, his staff and associated team members in the early stages of the Quiescent Period Prediction project. This report is dedicated to the memory of NAWCAD's late Chief Engineer Robert Ernst, and to the late Dr. Peter J F O'Reilly and Dean Carico, fathers of the Dynamic Interface discipline.

REFERENCES

1. BELMONT M.R., CHRISTMAS J., DANNENBERG J., DUNCAN J., DUNCAN J.M and FERRIER B. An Examination of the Feasibility of Linear Deterministic Sea Wave Prediction in Multi-Directional Seas Using Wave Profiling Radar: Theory, Simulation and Sea Trials. *Journal of Atmospheric and Oceanic Technology*, Vol. 31, **No. 7**, pp1601-1614, 2014.

2. CROSSLAND, P., FERRIER, B., APPLEBEE, TR. (2004). Providing Decision Making Support to Reduce Operator Workload for Shipboard Air Operations. Royal Institute of Naval Architects. London.

3. FERRIER, B., ERNST, R. (2019). Resolving the conflict between ship design and UAV Launch and Recovery deck limits; the development of enhanced dynamic interface study tests. American Helicopter Society Specialist Conference. Mesa.

4. APPLEBEE, TR., BAITIS, AE., MEYERS, W.G. (1981) "User's Manual for the Standard Ship Motion Program, SMP" David W. Taylor Naval Ship R&D Centre Report DTNSRDC/SPD-0936-01.

5. BLACKWELL, J AND FEIK, R. A (1988). "A Mathematical Model of the On-Deck Helicopter/Ship Dynamic Interface (U)". Aerodynamics Technical Memorandum 405. Aeronautical Research Laboratory. Melbourne.

6. FERRIER, B. (2008). MoD Quiescent Period Prediction LPD Software Testing Programme. UK080229-1. Bristol.

7. MORRIS, E.L., ZIENKIEWICZ, H.K., BELMONT, M. R (1998). Short term Forecasting of the Sea Surface Shape. *International Journal of Shipbuilding Prog.* **Vol. 45**, *No. 444*. Pp 381-400. Amsterdam.

8. FERRIER, B (2013). Deck Motion Characterisation Report. *LPD NSRS/QPP Operations*. UKSEA130219. Bristol.

8. BIOGRAPHIES

Michael Belmont, BSc, PhD, hold a chair at the University of Exeter UK, where he heads the Marine Dynamics group. He pioneered the new discipline of Deterministic Sea Wave Prediction which allows the short-term estimation of the details of wave driven vessel motion. This makes possible the prediction of quiescent intervals in the motion of naval platforms allowing the safe extension of the range of sea states for the launch and recovery of air, sea surface and sub-surface craft. Recently the same principals have been applied to the prediction of air-wakes over vessels. The goal of his research is the ability to allow the real time deterministic prediction of all the major environmental factors affecting maritime launch and recovery operations.

Jacqueline Christmas is an Associate Professor of Machine Learning in the Computer Science Department at the University of Exeter, UK. After many years working in industry, she obtained a PhD in the field of statistical pattern recognition in 2011. The main focus of her work is the development of machine learning models for remote sensing, with an emphasis on explain ability and measures of uncertainty, especially for potentially infinite sets of real-time observations, such as are collected by long-term, continuously operating sensors. Her particular interest is in the use of inexpensive, off-the-shelf sensors, such as video, radar and still cameras, to remotely sense difficult dynamic environments in real time. Since 2012 this work has been directed towards the measurement and prediction of sea waves, based on physical oceanographic theory.

Mr. Larry Fedrick FRINA MIET completed a fully indentured apprenticeship in MV/HV switchgear design in 1979 and embarked on a career in the construction industry specialising in power generation and distribution, Renewables, system testing and asset management before joining the UK MOD assigned as Royal Fleet Auxiliary heavy electrical Engineer in 2007, conducting numerous overseas assignments, then was appointed Fleet Electrical Equipment Authority for rotating electrical machinery in 2012 before being selected as Team Leader/Eng. Mgr. for Naval Design Partnering and Simulation Based Acquisition projects in 2016 and 2018 respectively. He leads the surface Warship platform pre-concept and concept design and Fleet in-service support design teams, simulation, and modelling projects in conjunction with UK and overseas stakeholders and industry partners for DE&S Ships Acquisition. Military experience includes 20 years RAF reserve forces as a trained fixed wing pilot/instructor.

Bernard Ferrier, PhD, FRINA, FRAeS is a Systems Engineering and Technology Assistant (SETA) in NAVAIR PMA-266 with the primary task of integrating and deployment of Fire Scout MQ-8C and its technologies and defining the operational limitations of the deck environment. Ferrier holds an honourary Professor position at the University of Exeter's College of Engineering. He is the Engineering Head of Hoffman Engineering's Aircraft-Ship Dynamic Interface Lab in Fairfax, VA. Ferrier led the CL227 UAV dynamic interface program at Bombardier in Montréal. Ferrier was a rotor dynamist and project leader of the dynamic interface program of the marinised AH-64 (Apache) and LHX programs at the McDonnell Douglas Helicopter Company. He received his last doctorate in helicopter/ship interface engineering at the École Polytechnique de Montréal (Québec) Canada.

Commander Robert Taylor is the RN Chief Naval Test Advisor and is responsible for the delivery of all Fleet Air Arm Test and Evaluation activities. He trained as a Lynx observer and has completed numerous front line, training, and trials assignments over a 30-year career. His early career centred around flying and included deployments to the Adriatic, the Baltic, and the Falklands, this was interspersed with training assignments on the Lynx Mk3 helicopter and its simulators. It was at this stage he began a career in test flying completing the Aerosystems Course and tours at the Lynx Operational Evaluation Unt and at Boscombe Down. Command tours include as a Lynx Flight Commander, Command of 700W Sqn, the first RN Wildcat Sqn, and of 744NAS, the Mission Systems and Weapon Integration Sqn based at Boscombe Down.Staff roles have centred around NCHQ, predominantly centred round assurance and ship-air integration. He has also completed an operational tour based in US Central Command, working closely with the US Joint Operations Centre and the Coalition support to the Middle East. He has completed an MBA in his spare time and an in-service MSc in Aerospace Systems. He is married to Joanna, a doctor, and they have 2 sons. He is a keen sportsman, keeping fit and enjoying golf and alpine skiing, and is an amateur woodworker and beekeeper.

THE ZOOM YEARS: UCL MSC STUDENT WARSHIP DESIGNS THROUGH THE PANDEMIC

R J Pawling, UCL Department of Mechanical Engineering, UK

SUMMARY

The global pandemic placed enormous strain on teaching, both for staff and students. This was most keenly felt in group projects, such as the UCL postgraduate ship design exercise, the 2019-20 cohort facing intense disruption as they switched "modes" half-way through the year. Despite this, the students were able to develop some innovative designs to meet challenging requirements. This paper provides an overview of the main themes for the last two years, both in the requirements set by staff and particularly the design response, which has included aspects such as optional crewing and future fuels, concluding an update on recent and future developments in the course that are most relevant to naval design.

1. INTRODUCTION: THE UCL MSC SHIP DESIGN EXERCISE

The UCL Department of Mechanical Engineering offers ship design teaching at undergraduate level (as part of a marine minor) and postgraduate level (on Naval Architecture [1] or Marine Engineering [2] MSc courses). The latter is the modern-day descendant of the UK MoD RCNC training courses that were run at Greenwich until 1967. The MSc Ship Design Exercise (SDX) is a 45 credit (450 hours individual engagement) group design module, where students work in teams of 3-5 to develop designs to meet challenging requirements.

An example User Requirements Document (URD) is provided in Appendix A. There have been additions to this format reflecting aspects of the latest (fourth) edition of the Engineering Council AHEP, the basis on which the courses are accredited, [3] including environmental and equality / inclusivity requirements.

3. THE IMPACT OF THE COVID-19 PANDEMIC ON TEACHING

The UCL SDX has been described in previously, [4] and to RINA Warships in 2019 [5] so this paper will not cover the procedural details. However it is worthwhile briefly discussing the impact of the COVID-19 pandemic. Whilst the possibility of a localised issue such as a class-wide outbreak of seasonal flu had been long considered a possibility, lockdowns and a move to all-remote teaching represented additional challenges.

Whilst many teaching staff saw, in move to remote teaching, an opportunity to incorporate more modern and sophisticated teaching technologies, the reality was far more brute-force. Research was very much deprioritised in favour of generating new on-line material such as pre-recorded and interactive material, running many more interactive sessions, rewriting exams for remote assessment etc. Fortunately for the teaching staff and students, UCL directed additional resources at the administrative and technical side. This was particularly important in the SDX as it included remote access tools for students to use software such as Paramarine and MaxSurf from their personal machines.

The 2019-20 cohort had the advantage that they had met each other prior to lockdown and so were able to maintain their design teams when working remotely. For the 2020-21 cohort, this was more of a challenge, as teaching was still remote and, in many cases, students did not meet one another face-to-face until the summer period, after the taught component.

One key response from the course staff was to increase active outreach to the students, as there were no "open office doors" for them to knock on. In the case of the SDX, this meant scheduled sessions with each group at roughly twice the frequency of a normal in-person year, along with the usual ad-hoc meetings via Skype etc. Additionally, guidance was issued to markers in both the 2019-20 and 2020-21 years, noting that the depth of some technical aspects would be limited. In the event, the students rose to the challenge and the resulting design reports were broadly in-line with previous years, both in terms of quality, innovation, and the inevitable "teachable moments" of any student project.

5. 2019-2020 DESIGNS

5.1 DESIGN REQUIREMENTS

The 2019-20 cohort was divided into six groups; four naval and two civil. The design requirements are summarised in Table 1, along with the composition of the design groups. Table 2 shows the budgets and class sizes associated with the requirements. Flexibility in number of ships is provided so that the students have to compare cost and capability options, with a costing model that includes shipyard learning over longer production runs.

Table 1: Design requirements for 2019-20

No.	Design Requirement	Outline	Group Size
1	AAW UxV Mothership	A vessel to provide local and area air defence against a range of projected high-speed above-water threats, using uncrewed vehicles (UxVs) and offboard systems.	3NA 1MAR
2	Future LXX	A common hull multi-purpose amphibious warfare mothership class	3NA 1MAR
3	Depot ship and associated large USV	A multi-vessel system consisting of a depot ship and associated large USVs (>20t and <200t displacement). The design of both the depot ship and the USV will be required. The USV need not be developed to the same detail as the depot ship, but the integration of the two as a system should be demonstrated.	3NA 1MAR
4	Upgradeable OPV	An Ocean Patrol Vessel that can be upgraded to a frigate part way through life by the addition of a parallel mid-body hull extension.	2NA 1MAR
5	Wind assisted trans-oceanic passenger liner	A wind assisted passenger liner for trans-Atlantic voyages, aimed at travellers wishing to reduce their carbon footprint	2NA 1MAR
6	Ultra-low emissions fast ferry	A fast ro-pax fast ferry for passengers and vehicles, with a secondary naval theatre transport role and an exceptionally low emissions requirement.	2NA 1MAR

Table 2: Budget allocations for the 2019-20 UCL ship design exercise

No.	Ship	Budget, Millions of Pounds (M£)	Number of Ships
1	AAW UxV Mothership	9000 M£ for class excluding UxVs	8 minimum
2	Future LXX	4000 M£ total	TBC by group
3	Depot ship and associated large USV	9000 M£ total including USVs (excl. attrition)	3 motherships and 8 large ASW USVs minimum
4	Upgradeable OPV	100 M£ each for OPV, 100 M£ each for upgrade (on average)	Target is 16
5	Wind assisted trans-oceanic passenger liner	Demonstrate cost effectiveness - should at least provide cheaper tickets than a luxury cruise liner	Equivalent to aircraft: 100-400 passengers
6	Ultra-low emissions fast ferry	450 M£ for class	3 is preferred option

5.2 AAW UXV MOTHERSHIP

This design was essentially a projected replacement for the RN Type 45 Destroyer, having the same performance requirement as the previous years "Future Missile Fighter-Heavy" described in Pawling [5] only with the proviso that the capability should come from UXVs.

The students were provided with a range of surveillance and armed UAVs and USVs and could choose between them. The area Anti-Air Warfare (AAW) system selected used 9 x subsonic Short Take-Off Vertical Landing (STOVL) UAVs for surveillance, with a long-range silo-launched SAM for the primary effector. The 14 x 14m 5 tonne delta-winged AEW UAV, illustrated in Figure 1 originated in a 1990s MoD study.

The secondary AAW effector selected was 9 x tiltrotor VTOL UAV carrying air-to-air missiles, slower to respond than the SAM but useful for filling gaps and providing a flexible response, the tiltrotor being derived from the Bell V-247 Vigilant. The principal particulars of the AAW UXV mothership are summarised in Table 3 and the design is illustrated in Figure 2.

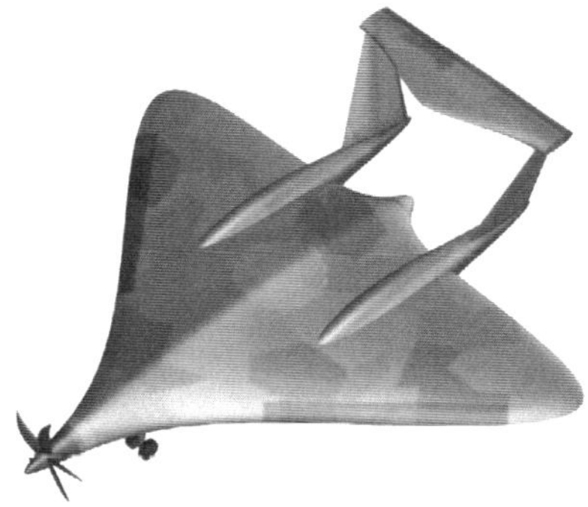

Figure 1: STOVL AEW UAV, showing turboprop propulsion and antennas in wing edges.

Table 3: AAW UXV mothership particulars

Displacement, light	8600te
Displacement, deep	13600te
Internal volume	55700m^3
Length, waterline	198m
Beam, waterline	22.3m
Draught, hull	5.5m
Depth, hull	18m
Speed, cruise	15 knots
Speed, max	25 knots
Complement	201
FoC cost	£1.28Bn

Being a UXV rather than UAV carrier, the mothership included two stern ramps for large USVs. These provided sea control and counter small-boat capabilities. The design was dominated by the comparatively large hangar internal to the hull, with fine subdivision required to avoid flooding of the hangar in damage scenarios. The second, deck edge aircraft lift shown in Figure 2 was accepted to only be usable in good weather.

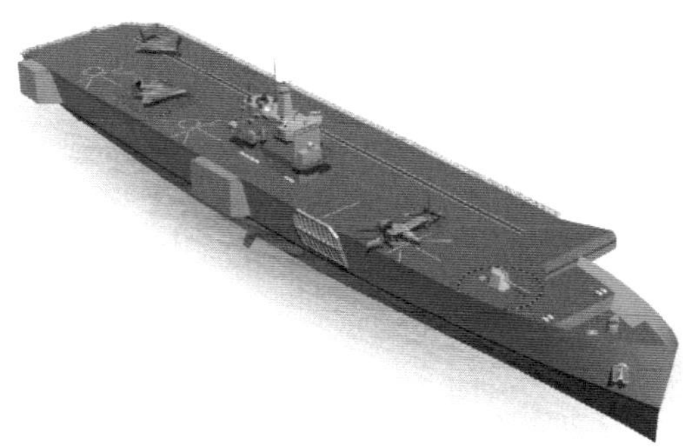

Figure 2: the AAW UXV mothership design

The propulsion architecture summarised in Table 4 and Figure 3 was relatively conventional: a CODLAG arrangement with four MTU diesel generators and a single MT-30 gas turbine. A battery-based energy storage system and PEM fuel cell were included to satisfy the requirement for an ultra low emissions mode in ports and coastal areas.

Table 4: Propulsion system principal particulars

CODLAG architecture
2x3MW PMS motors for cruise
1xRR MT30 GTs (36MW) for sprint
4x3MW MTU 20V4000 DS3250
1x2.2MWh LiOH Battery
600V Main AC distribution
1x0.5MW PEM Fuel Cell

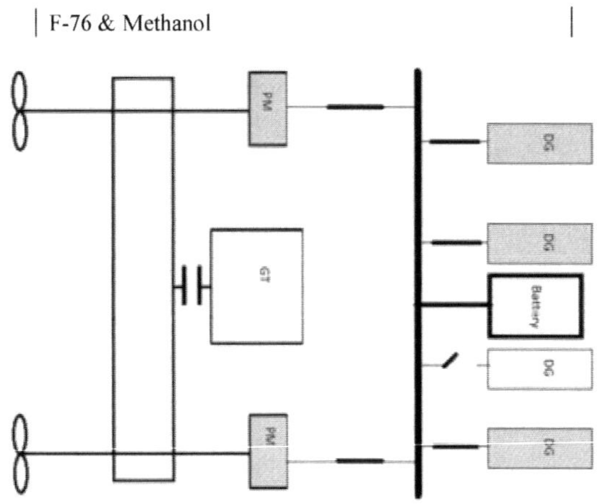

Figure 3: Propulsion system line diagram

5.3 FUTURE LXX

This design requirement was for a fleet of multiple ships, variants of a common hull, to fulfil a range of roles in an amphibious operation, including; the deployment of a battalion-sized marine force; provision of tactical close air support; medical facility; strategic sealift. The number and type of variants of the common hull was a decision for the students, based on operational analysis.

The resulting design was a Landing Platform Dock (LPD) with vehicle deck and large hangar in the superstructure to support deployment of a range of UAVs. The LXX principal particulars are summarised in Table 5 and the design is illustrated in Figure 4. The students aimed to incorporate adaptability in the design, with the well dock able to take various combinations of landing craft, such as two US Navy LCAC, four QinetiQ PASCAT or four BMT Caimain-60. The vehicle deck was designed with widened lanes to accommodate future armoured vehicles and equated to 700 commercial lane-metres.

Typical for an amphibious mothership, the LXX design was equipped with basic self-defence capabilities in the form of small calibre guns, CIWS and point defence missiles.

Table 5: LXX principal particulars

Displacement, light	12500te
Displacement, deep	20400te
Internal volume	71300m^3
Length, waterline	183m
Beam, waterline	25.5m
Draught, hull	6.72m
Depth, hull	18.1m
Speed, cruise	15 knots
Speed, max	20 knots
Complement	109
EMF troops	678
FoC cost	£645M

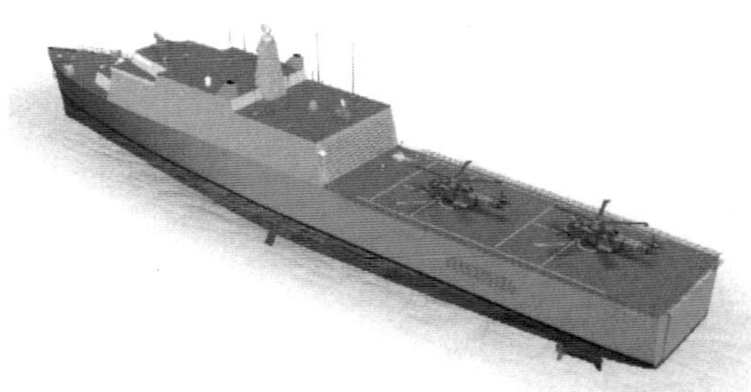

Figure 4: LXX baseline design

Like the AAW UXV Mothership, the LXX team selected NATO F-76 as the primary fuel, with ammonia fuel included for low-emissions coastal operations. The IFEP power and propulsion system used a combination of dual-fuel diesel engines and batteries ("energy magazines") powering two 7.5MW propulsion pods. Figure 5 shows the distribution of power and propulsion machinery in the hull, with the tank arrangement shown in Figure 6 which clearly shows the two cylindrical ammonia tanks.

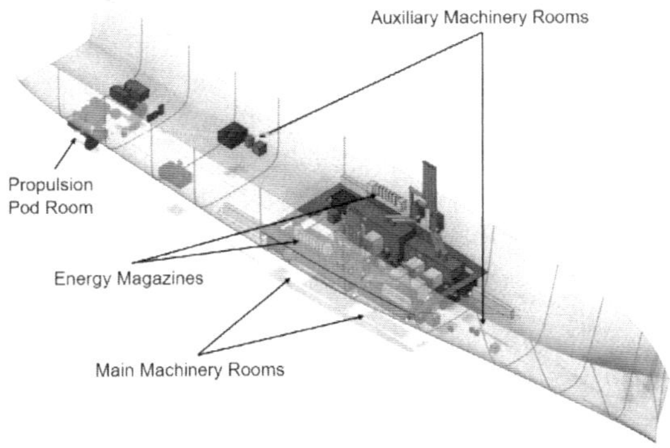

Figure 5: LXX power and propulsion systems

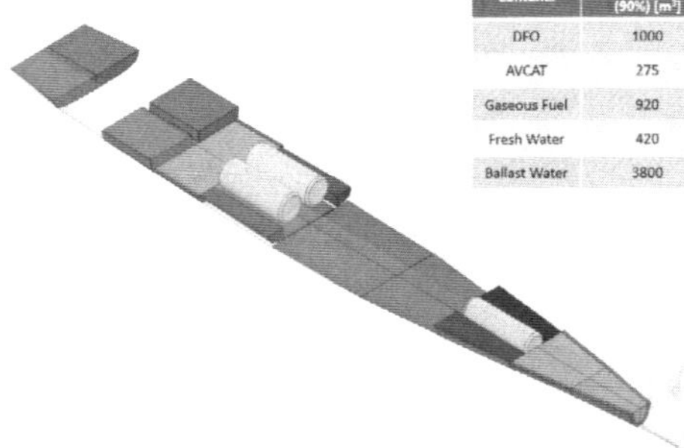

Contents	Working Capacity (90%) [m^3]	Colour
DFO	1000	
AVCAT	275	
Gaseous Fuel	920	
Fresh Water	420	
Ballast Water	3800	

Figure 6: LXX main tanks

Also apparent in Figure 6 is the fact that the ballast tanks are distributed along the length of the vessel. This was due to the provision of a proportionally long internal dock, for flexibility in the choice of ship to shore interconnector. This reduced the potential to use simple trim to flood the well dock, with parallel sinkage required to maintain overhead clearance at the stern entrance whilst also providing sufficient water depth at the internal beach.

5.4 DEPOT SHIP AND ASSOCIATED LARGE USV

Group 3 had a design requirement relatively novel for the UCL course: to design not just a mothership for existing or generic vehicles in the UCL database, but also design a bespoke large USV to meet the operational requirements. This requirement arose from UCL staff looking at current USVs and noting that those intended for deployment from motherships were generally based on existing small boat designs, with large USVs generally not intended for deployment from a mothership.

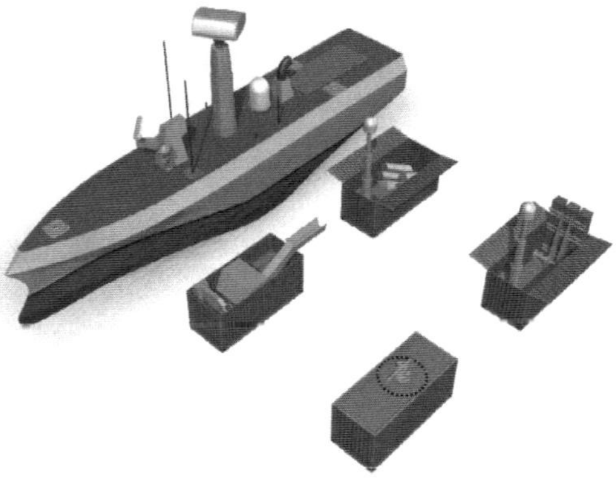

Figure 7: The bespoke USV, showing modules for ASW, ASVW, AAW and anti-small boat

The primary roles for the USVS were ASW; where they acted as towed-array "tugs"; AAW; there they provided a screening force armed with CAMM missiles; and ASVW; using light anti-ship missiles and group tactics. The requirement to tow large towed-array sonars, combined with a desire to keep length down to reduce the impact on the mothership design, led to the adoption of a small trawler hullform as the basis for the USV, with a large module bay accommodating combat systems in TEU-sized modules, as illustrated in Figure 7.

Table 6: USV depot ship and USV principal particulars

Mothership		**USV**	
Displacement, light	7300te	**Displacement, light**	130te
Displacement, deep	10650te	**Displacement, deep**	150te
Internal volume	45300m^3	**Internal volume**	
Length, waterline	146.6m	**Length, waterline**	25.67m
Beam, waterline	22.55m	**Beam, waterline**	5.90m
Draught, hull	5.66m	**Draught, hull**	2.35m
Depth, hull	14m	**Depth, hull**	5.5m
Speed, cruise	18 knots	**Speed, cruise**	15 knots
Speed, max	28 knots	**Speed, max**	25 knots
Complement	151+30 EMF	**Endurance @ 11knots**	100 hours
FoC cost	£630M	**Unit cost**	£20M (Max module cost £19M for ASW)

Although multihulls were considered for the mothership, a monohull was selected for development as the broad transom would allow a range of ocean interface options USVs too large to lift with a crane. The mothership was designed to carry a maximum of 6 bespoke USVs and 12 10-tonne USVs derived from existing types. A twin hangar was included for crewed helicopters and rotary wing UAVs. Figure 8 illustrates the mothership, including the side doors from the large superstructure mission bay containing the smaller USVs. Unsurprisingly, the mothership design was dominated by the USV stowage spaces and launch and recovery methods. Figure 9 shows the total volume of the ship occupied by USV and aircraft spaces. Whilst smaller USVs were launched and recovered by side davits, the large USVs used a movable stern ramp, where a cradle-type structure would be moved on tilting rail sections to allow launch and recovery.

Figure 8: The mothership design with two USVs

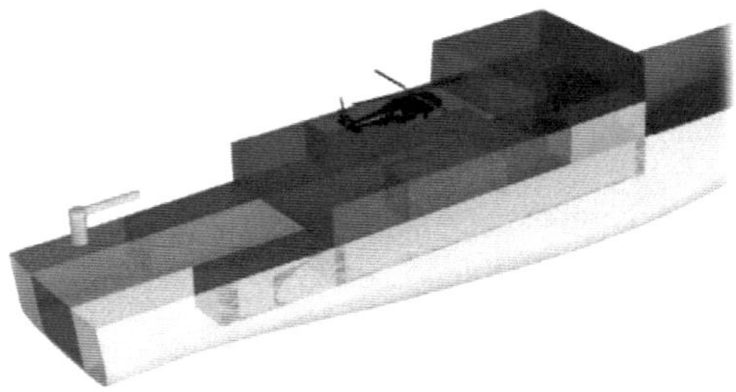

Figure 9: USV and aviation spaces in the mothership.

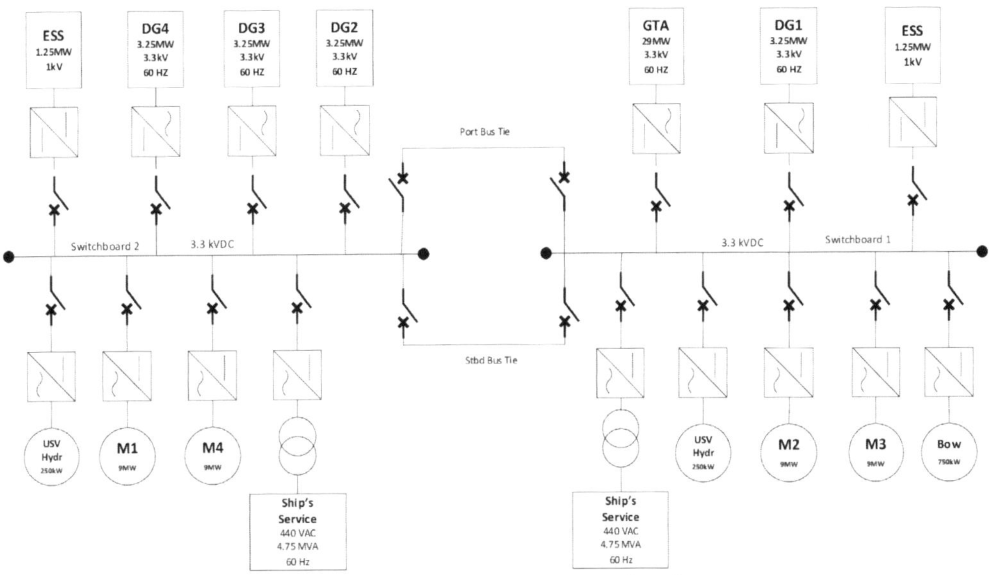

Figure 10: Mothership electrical system line diagram.

With a highly varied operational profile (including long periods at slow speeds for USV launch and recovery) and large varying loads for USV handling systems, an all-electric propulsion system was selected for the mothership, using DC distribution. The mothership power and propulsion system line diagram is shown in Figure 10. The USV also employed a diesel-electric propulsion system, however in that case the driver was arrangement, in that electrification allowed the various items of propulsion equipment to be arranged in a short hull.

5.5 UPGRADEABLE OPV

This design requirement was developed by the UCL staff as an alternative take on the broad requirements for Type 31. Available information at the time indicated a design solution that would have capabilities similar to a high-end OPV at the start of life, with later additions likely on the same hull. The students were tasked with developing a design which could be modified mid-life with a hull extension, so changing an OPV into a frigate. Figure 11 illustrates the design in OPV and post-extension versions.

Figure 11: Group 4 design in OPV (rear) and frigate (front) configurations

The 30m parallel mid-body contained additional machinery, accommodation and an 8-cell strike-length VLS for anti-ship and ASW missiles, with quad-packed CAMM for self-defence. The upperdeck space allowed for two Phalanx CIWS and the datalinks associated with the missile systems.

Table 7: Principal particulars in OPV and typical frigate configurations

OPV		**Frigate**	
Displacement, light	1880te	**Displacement, light**	2680te
Displacement, deep	2370te	**Displacement, deep**	3640te
Internal volume	8160m^3	**Internal volume**	11160m^3
Length, waterline	91.4m	**Length, waterline**	121.4m
Beam, waterline	12.8m	**Beam, waterline**	12.8m
Draught, hull	4.0m	**Draught, hull**	3.8m
Speed, cruise	12 knots	**Depth, hull**	9.0m
Speed, max	20 knots	**Speed, max**	27 knots
Complement	40	**Complement**	150
FoC cost	£138M	**Mid-body cost**	£102M

Whilst items such as VLS or additional accommodation could be included in a midships extension, the power, propulsion and hotel systems were more complicated. Group 4 adopted a CODLAG architecture, where the OPV would be built with motors driving two shafts via gearboxes. The midship extension would add an additional diesel generator (to meet increased hotel load) and a gas turbine with a splitter gearbox connected to input spurs included on the gearboxes. Figure 12 shows the OPV/FFG power and propulsion system, with dot-dash lines around those items added in the refit. The small size of the vessel ruled out larger gas turbines, with a power of 7-10MW required; with the Siemens SGT-200 (7.6MW) selected.

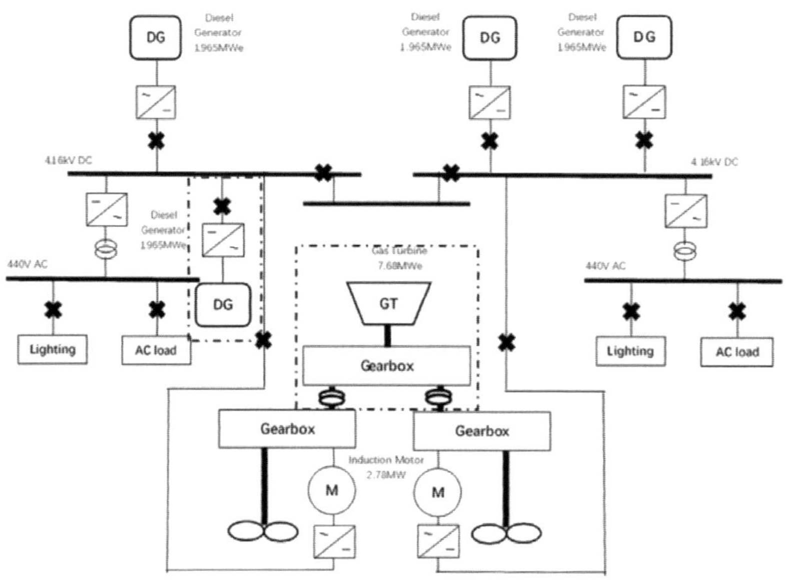

Figure 12: OPV/FFG power and propulsion system line diagram

As noted, this design requirement was developed to investigate this concept as an alternative to the Type 31, and it is worthwhile noting that it was found to not be as desirable an idea compared to simply building a large ship to start with. The students themselves determined this (without prompting from the staff), and this underscores the fact that, in the UCL design exercise marking scheme, students who determine that the resulting design – or even the overall concept – is flawed will be rewarded for identifying this and explaining why. Of course, it can be a challenge to convince students that they can in fact pass the module with a design that fails the requirements!

6. 2020-21 DESIGNS

6.1 DESIGN REQUIREMENTS

The 2020-21 cohort was divided into seven groups; five naval and two civil. The design requirements are summarised in Table 8, along with the composition of the design groups. Table 9 shows the budgets and class sizes associated with the requirements.

Table 8: Design requirements for 2020-21

No.	Design Requirement	Outline	Group Size
1	Ultra-Stealthy Surface Combatant	A blue-water escort with exceptionally low signatures.	2NA1MAR
2	Persistence Carrier	A low-cost aircraft carrier designed to offer a range of capabilities for long-duration, low-intensity missions.	2NA 1MAR
3	Multipurpose Logistics Vessel	A naval auxiliary providing solid stores replenishment and logistical support to expeditionary warfare and disaster relief.	2NA 1MAR
4	Optionally Crewed Territorial Defence Vessel	A small, high speed vessel for patrol and defence of home waters against a numerically superior adversary.	2NA 1MAR
5	Common Hull Surface Combatant	A medium-sized surface combatant that makes use of design modularisation to allow completion in a variety of roles.	2NA 2MAR
6	Green and Distanced Cruising	A cruise liner with low environmental impact and a high level of bio-security.	2NA 1MAR
7	Coastal Freight System	A Ro-Ro freight vessel for short sea routes, exploiting automation and offboard vehicles.	2NA 1MAR

Table 9: Budget allocations for the 2020-21 UCL ship design exercise

No.	Ship	Budget, Millions of Pounds (M£)	Number of Ships
1	Ultra-Stealthy Surface Combatant	£3Bn	10 minimum
2	Persistence Carrier	£1500M not including aircraft	3
3	Multipurpose Logistics Vessel	£1600M	3-5
4	Optionally Crewed Territorial Defence Vessel	£1300M	5 is minimum
5	Common Hull Surface Combatant	£10Bn over three batches	4 x 3
6	Green and Distanced Cruising	£1500M	4
7	Coastal Freight System	£700M not including aircraft	8-12

6.2 ULTRA-STEALTHY SURFACE COMBATANT (USSC)

This requirement was for a relatively conventional escort frigate with the primary role of ASW but able to contribute towards task group AAW and ASVW, with the reduction of signatures to the maximum extent. The combat system is summarised in Table 10.

Table 10: USSC combat systems

Sensors and Electronics	
	SAMPSON multifunction radar
	Bow, towed and dipping sonars
	Frigate comms fit
	ESM
Weapons	
	1 x 30mm gun
	2 x 12.7mm guns
Launchers	
	4 x 8 strike length VLS
	2 x 4 cell light VLS (CAMM equivalent)
	Decoy launchers
	127mm anti-torpedo torpedoes
Daughter Craft	
	2 x 11m RHIB or USV

| 1 x 5 tonne helicopter |

Figure 13 is an illustration of the overall design, showing the clean superstructure and extensive bulwarks to hide upperdeck equipment. The primary functional difference between this vessel and a contemporary ASW frigate such as Type 23 was that the students opted to use lighter offboard assets, such as Lynx Wildcat, for submarine detection, with prosecution of targets via ship-launched ASW missiles. The principal particulars of the final design are summarised in Table 11.

Table 11: USSC principal particulars

Displacement, light	3580te
Displacement, deep	4335te
Internal volume	17580m^3
Length, waterline	125m
Beam, waterline	14m
Draught, hull	4.3m
Depth, hull	9.05m
Speed, cruise	18 knots
Speed, max	29 knots
Complement	130
Average unit cost	£300M

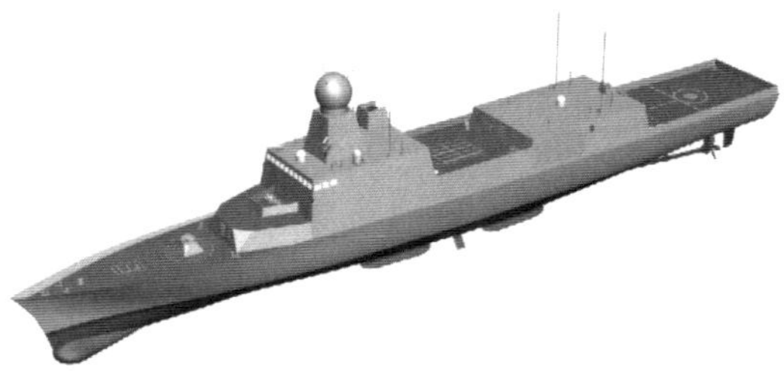

Figure 13: Ultra-stealthy surface combatant

The 2020-21 cohort had a greater emphasis on future fuels and emissions reduction in their URDs and this was reflected in the Ultra Stealthy Surface Combatant by the use of methanol fuel and high-temperature proton exchange membrane (HT-PEM) fuel cells cruise propulsion and hotel load supply via a DC grid. A single LM2500 gas turbine driving the shafts via a splitter gearbox was included for sprint speeds. As shown in Figure 14, the propulsion motors selected were high-temperature superconducting types, requiring a liquid nitrogen plant for cooling.

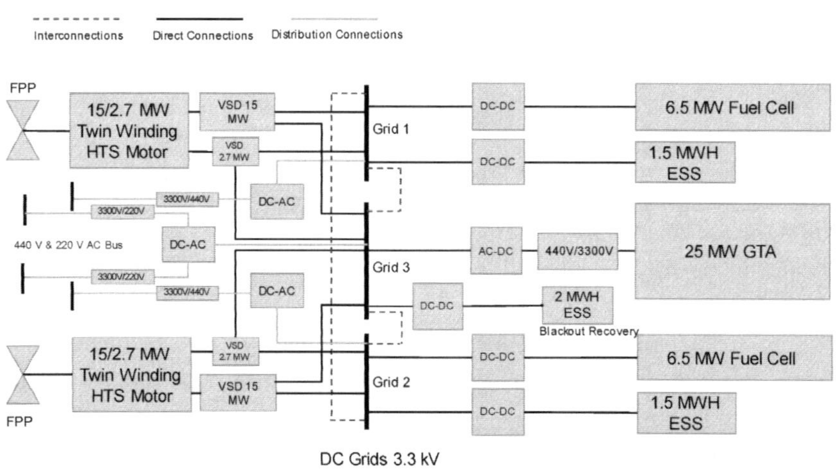

Figure 14: USSC power and propulsion system line diagram

The stealth requirement was addressed with several different design features:

Radar signature:

- Reduction of upperdeck clutter and use of bulwarks
- Overall geometry

Infra-red:

- Use of fuel cells for cruise speeds
- Shielding of exhausts

Underwater noise:

- Use of fuel cells for cruise speeds
- Application of rubber isolating tiles to hull around machinery spaces

The RCS was assessed using the CADRCS implementation within QinetiQ's Paramarine software, allowing identification of RCS "hot spots" and comparison of the overall signature against generic values provided for reference. An example of the resulting RCS is shown in Figure 15.

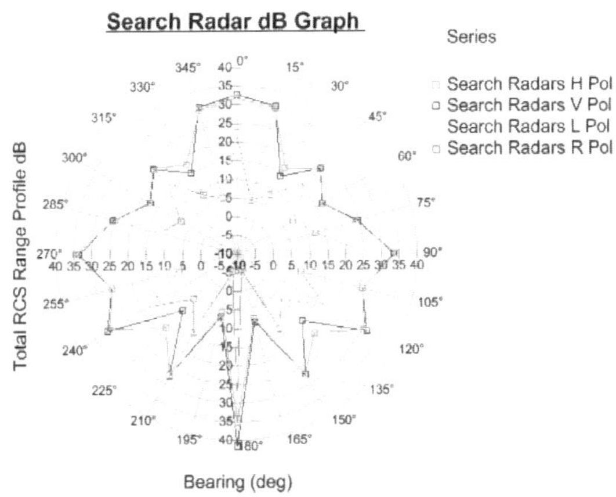

Figure 15: Group 1 RCS analysis

6.3 PERSISTENCE CARRIER

Requirement 2 called for an aircraft carrier intended for long duration, low tempo operations using a diverse range of aircraft. Light displacement was capped at 40,000 tonnes. The solution developed by the students is illustrated in Figure 16, with the principal particulars shown in Table 12, which shows how the limiting factor in the design was the budget rather than displacement cap.

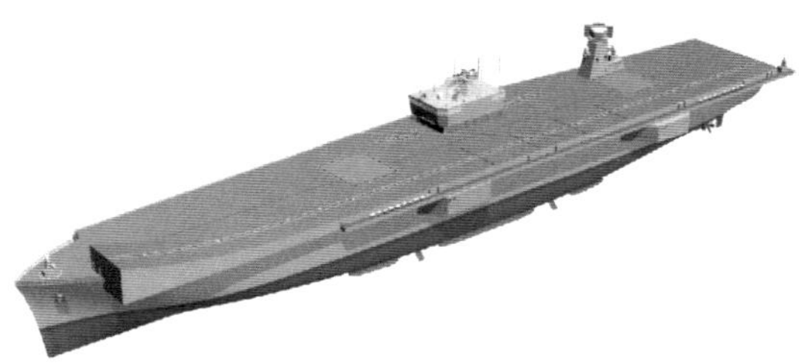

Figure 16: Persistence carrier overview

Table 12: Persistence carrier principal particulars

Displacement, light	17660te
Displacement, deep	27470te
Internal volume	137300m^3
Length, waterline	240.1m
Beam, waterline	30.5m
Beam, overall	48.0m
Draught, hull	6.7m
Speed, cruise	15 knots
Speed, max	25 knots
Complement	850+220 EMF

FoC cost	£517M

The persistence carrier design includes the typical point defence SAMs, CIWS and small guns. A typical air wing composition is shown in Table 13. The baseline design was for the larger aircraft (F-35 and AEW UAV) to be STOVL whilst smaller UCAV such as the Predator and X-47B sized machines would take off using a full-length flight deck run and bow ramp, but recover with arrestor gear. This gear was again budget limited and not sized for crewed fighter aircraft. Space, weight and power allowances were included in the design for a later refit with an electromagnetic catapult for use with the smaller UCAVs.

Table 13: Persistence carrier typical air wing composition

Aircraft Type	**Number Carried**
F-35B	6
AEW UAV	6
Predator B / Mariner	2
Strike UCAV	2
Wildcat	4
Merlin	2

The power and propulsion system for the persistence carrier adopted an IFEP architecture with DC distribution, with a total of eight diesel generators and a substantial 30MWhr lithium-ion battery bank for backup and zero-emissions operation. The vessel was sized to accommodate the larger tanks required for a mid-life refit to use renewably sourced methanol fuel in place of F-76.

Survivability was a challenge for a small aircraft carrier with an internal hangar of 7.5m height. Iteration of the bulkhead locations resulted in a design where the hangar deck was not submerged in any three-compartment damage case, however it was not possible to incorporate any longitudinal access below the hangar.

The susceptibility aspect of survivability was addressed with reduction of the above water RCS, with particular attention paid to the superstructure that would be visible "over the horizon". Figure 17 shows the "before" and "after" RCS calculations for a broadside view, with red areas providing the strongest radar returns.

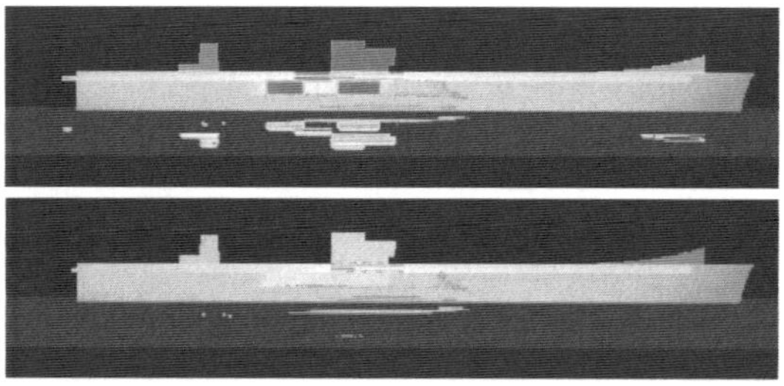

Figure 17: RCS calculations for a broadside view, showing reductions in superstructure contributions from the baseline (top) to the final design (bottom)

6.4 MULTIPURPOSE LOGISTICS VESSEL (MPLV)

The MPLV requirement was for a solid stores replenishment ship with a secondary capability for "opportunistic" refuelling of smaller vessels such as OPVs or USVs. The special study was that the MPLV should be able to support "continuous resupply" concepts, where UAVs and USVs are used to move supplies in a continuous flow in addition to the conventional major alongside replenishment activities. The design is illustrated in Figure 18 and the principal particulars are given in Table 14.

Figure 18: The MPLV

Table 14: MPLV Principal particulars

Displacement, light	5048te
Displacement, deep	13302te
Internal volume	12978m^3
Length, waterline	127m
Beam, waterline	27.2m
Draught, hull	6.9m
Depth, hull	13m
Speed, cruise	15 knots
Speed, max	20 knots
Complement	50 core + 100 EMF
FoC cost	£450M

Table 14 shows the typically very high variable load of a supply ship. Additionally, the low length to beam ratio should be noted – the student group found the design driven to a large beam due to initial estimates of KG.

Combat systems were typical of a naval auxiliary; a 16-cell point defence missile system, CIWS, small calibre guns and decoys. Consideration of USV vs UAVs for the continuous supply concept led to the selection of UAVs and cargo helicopters, driven by the greater ship impact of stern ramps or a dock and the flexibility of the aviation option. The design used a generic cargo UAV based on the uncrewed Kaman K-MAX variant used by the USMC.

The MPLV power and propulsion system architecture selected was a relatively conventional twin-shaft CODLOD arrangement, with each shaft driven by either a 10.8MW engine or a 3.3MW gearbox mounted motor. Four 3.2MW diesel generators provided hotel and cruise propulsion power, with 2MWhr of battery storage. The drive motors were configured to operate as generators. This Power Take In / Power Take Out (PTI/PTO) solution was adopted in several student designs as it offers flexibility and survivability.

A profile view of the MPLV is shown in Figure 19, with the solid stores spaces highlighted in orange. These spaces took up the majority of Nos 2 and 3 deck in the hull, with large lifts to the upperdeck leading to the large gap amidships. RASCO in this design being in the after part of the forward superstructure.

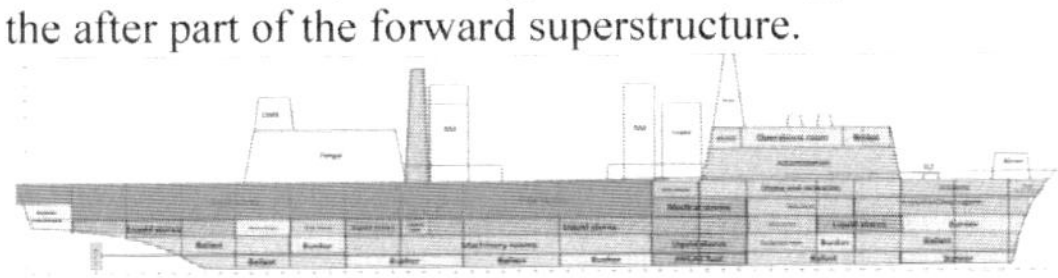

Figure 19: MPLV profile with solid stores spaces highlighted

6.5 OPTIONALLY CREWED TERRITORIAL DEFENCE VESSEL (OCTDV)

The complete requirement issued to the students for this vessel is included as an example in Appendix A. It called for a vessel of less than 2000 tonnes standard displacement that could be operated in an uncrewed mode as part of a group of vessels, with an emphasis on the defence of home waters. The resulting design is illustrated in Figure 20 and the principal particulars are given in Table 15.

The basic combat systems were typical of a small corvette and are summarised in Table 16. Notably a permanent hangar was omitted in favour of stowage for modules that included additional accommodation for personnel-intensive roles such as the command of a strike group, or additional weapons for the semi-autonomous "wingmen" in such a formation.

Figure 20: OCTDV overview with ASM modules

Table 15: Territorial defence vessel principal particulars

Displacement, light	1274te
Displacement, deep	1611te
Length, waterline	85.5m
Beam, waterline	9.9m
Draught, hull	3.7m
Depth, hull	8m
Speed, cruise	15 knots
Speed, max	28 knots
Complement	8 + 22 modular
FoC cost	£183

Table 16: OCTDV combat systems

Sensors and electronics	
	3D radar (e.g. ARTISAN)
	Hull mounted sonar
	Small ship comms with enhanced datalinks
Weapons	
	76mm gun
	2 x 30mm guns
	Phalanx CIWS
Launchers	
	8 cell strike length VLS
	Decoy launchers
	2 10 x 3 m modular spaces
	Davits for 2 x 7m boats

The OCTDV had a particularly onerous emissions reduction requirement, as it would spend most of its operational time patrolling home waters, which were likely to be Emissions Control Areas (ECAs). The power and propulsion system solution was a combined fuel-cell or gas turbine architecture, with fuel cells fuelled by hydrogen used for peace time patrol, whilst the high-speed operations expected in most defensive wartime scenarios would use gas turbines fuelled by diesel. Figure 21 shows the overall power and propulsion system line diagram. Key to this concept was the use of two 1MW motor-generators. In peacetime they would be used as motors, powered by the hydrogen fuel cells. In wartime they would however function as generators, driven by the four 4.6MW gas turbines.

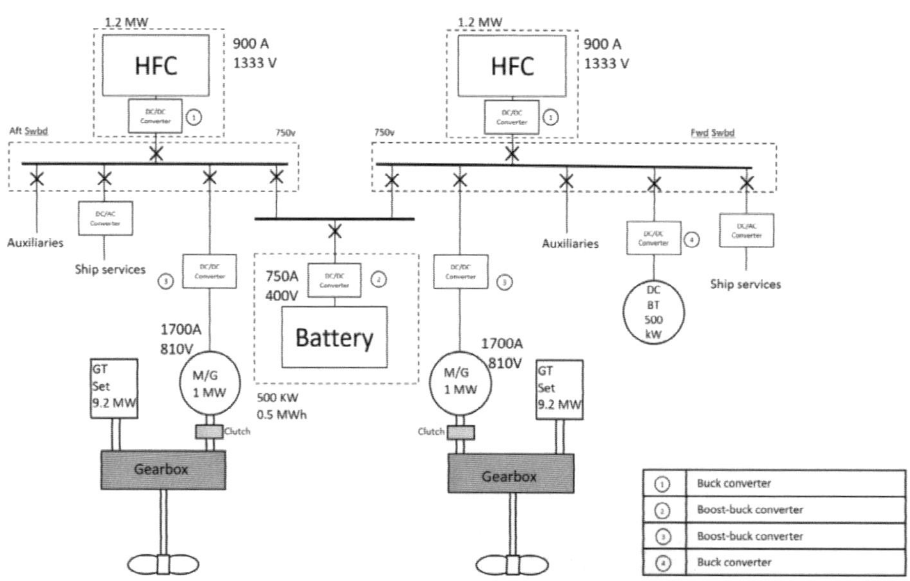

Figure 21: OCTDV power and propulsion system line diagram

The use of hydrogen fuel, stored in liquid form, had a significant impact on the layout and overall sizing. Figure 22 shows a profile view of the OCTDV, highlighting the fuel and machinery spaces. The cylindrical geometry of the hydrogen tanks led to them being higher in the ship than diesel tanks, and their weight produced a ship with a density much greater than conventional warships, over 0.3tonnes/m^3 overall.

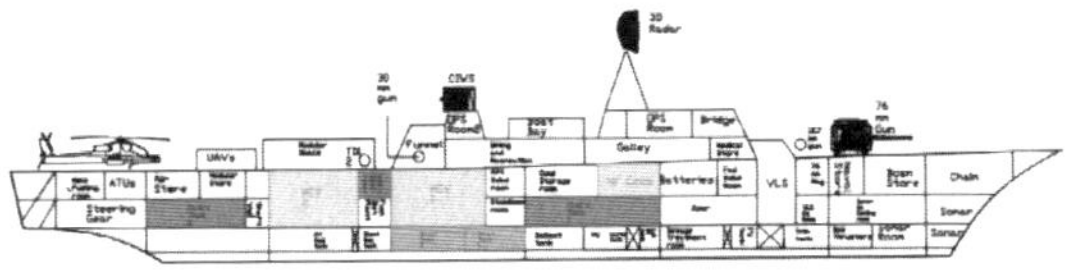

Figure 22: Profile view showing hydrogen tanks (blue), diesel tanks (orange) and machinery spaces (yellow)

An additional aspect of novelty in the design was that the hull was intended to be largely unoccupied, with the crew and operational spaces in the superstructure. Hydrogen gas and all machinery spaces would be unattended with an inert atmosphere of nitrogen gas during normal operation.

6.6 COMMON HULL SURFACE COMBATANT (CHSC)

2020-21 featured a common-hull design, but for this year it was a surface combatant. The concept of using a common hull for radically different roles, such as AAW and ASW, is not a new one, and has historically proven difficult [6]. For Group 5, the challenge was to provide a design that did not exceed 6500 tonnes standard displacement, which could be completed in AAW, ASW or ASVW / land attack variants. The resulting design is illustrated in Figure 23, with the principal particulars given in Table 17.

Figure 23: Common Hull Surface Combatant overview

Table 17: Principal particulars

Displacement, light	5400te
Displacement, deep	8400te
Internal volume	25500m^3
Length, waterline	155.2m
Beam, waterline	17.9m
Draught, hull	6.2m
Depth, hull	11.8m
Speed, cruise	15 knots
Speed, max	27 knots
Complement	130

The main capability differences between variants are summarised in Table 18. It should be noted that aviation and USVs featured heavily in all options. In the AAW variant, deployed craft were seen as providing early warning capability with USVs providing the option for sophisticated offboard electronic warfare, whilst high-speed hydrofoil USVs were selected to allow for rapid response to the small boat threat.

Table 18: High-level capability differences between variants

	AAW Variant	**ASW Variant**	**ASVW / Land Attack Variant**
Sprint speed	27 knots	27 knots	24 knots
Main gun	76mm	76mm	155mm
VLS cells	48	40	48
Key sensors	MFR + LRR	Bow and towed array sonar	MFR
Helicopters	2 x 15 tonne	2 x 15 tonne	2 x 15 tonne
USVs	4 x 10m USV	4 x 10m USV	4 x 10m USV
FoC UPC	£716M	£671M	£711M

The requirement to build three different variants had an impact on the general arrangement. The medium calibre gun and VLS are modular spaces, designed for the maximum capability but not fully occupied in all variants. More significantly, the entirety of the superstructure aft of the bridge was composed of USV and module bays and aviation. This concept allowed for additional electronics spaces needed for the AAW variant to be easily added.

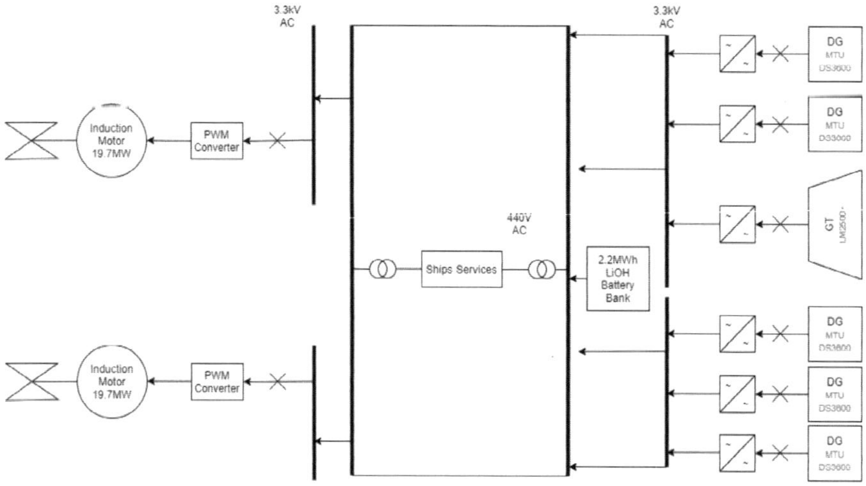

Figure 24: Power and propulsion system for the Common Hull Surface Combatant

An IFEP configuration with 3.3kV AC distribution was selected for the CHSC. Pime movers were five MTU diesel generators and one LM2500+ driving an alternator, with a 202MWhr LiOH battery for backup and zero-emissions modes. The power and propulsion system line diagram is shown in Figure 24. To comply with future emissions reduction regulations, methanol was selected as the most favourable future fuel. The higher volume demands associated with methanol led to the majority of the double bottom and no 5 deck being deep tanks, with some tanks carried up to No 4 deck.

7. TRENDS AND FUTURE DEVELOPMENTS IN THE UCL SHIP DESIGN EXERCISE

7.1 DEVELOPING REQUIREMENTS

As was noted in the introduction, the structure of the URD shown in Appendix A has seen additions for the 2021-22 cohort to reflect aspects of AHEP-4 and some general trends in the exercise. These additions include:

Human Factors (HF)
A dedicated section on HF, incorporating diversity and equality issues such as changing crew demographics, access for the disabled etc.

Emissions
This has been elevated to its own section, with both air emissions and also underwater emissions, including noise for commercial vessels.

Marine Engineering
Requirements have been more clearly separated between the power and propulsion system (itself a term deliberately chosen to encourage a systems approach) and the auxiliaries.

Background and Context Statements
The URD is issued with a set of statements describing the projected future "world" and the future marine fuel / energy seascape has influenced both the URD and background for 2021-22. A common approach (dating at least to the author's undergraduate years) has been to declare that navies will either be able to disregard legislation, or that a convenient "drop-in" biofuel will become available. This approach is now being strongly disincentivised within the exercise (at least for peacetime operations), although it is not explicitly ruled out.

7.2 ANALYSIS AND TOOLS

As more technical areas are added to the requirements document, so it becomes more important to ensure that analysis methods and tools are available, and that the marking scheme allows for differentiation. Modern students can be highly focussed on how to "tick the boxes" to maximise marks and the general guideline being adopted for the SDX is that, if the staff set a requirement, then we should be able to tell the students how we expect them to demonstrate compliance – be it by reference (e.g. IMO EEDI calculation methods) or by a UCL developed tool, or a taught approach e.g. structural design.

The latter has led to the development of several bespoke UCL tools, such as COBALT (COst Benefit AnaLysis Tool) which generates plots of cost vs. capability as shown in Figure 25; VANADIUM (Vulnerability ANAlysis for Design In the UCL Msc), an Excel-based early stage vulnerability analysis tool (Figure 26) or a range of wargaming tools and games for evaluating operational concepts and assigning capability benefit levels [7].

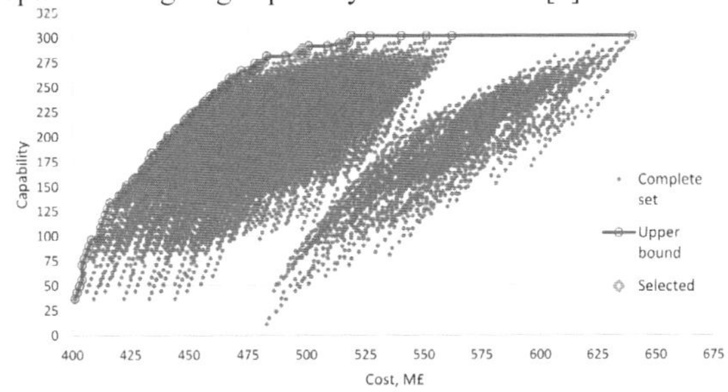

Figure 25: Example COBALT analysis output

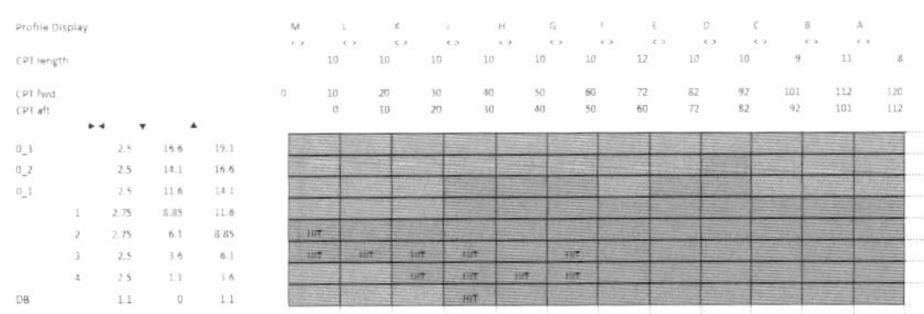

Figure 26: Example VANADIUM output showing critical hit locations for a propulsion system

9. ACKNOWLEDGEMENTS

The author would like to acknowledge the UCL MSc students of the 2019-20 and 2020-21 cohorts who worked so hard on their ship design under exceptional circumstances, providing the source material for this paper. The author would also like to acknowledge here UCL colleagues and external industry lecturers whose input is vital for the course.

10. REFERENCES

1. "MECH0069 Ship Design Exercise for Naval Architects", https://www.ucl.ac.uk/mechanical-engineering/study/postgraduate-taught/naval-architecture-msc

2. "MECH0084 Ship Design Exercise for Marine Engineers", https://www.ucl.ac.uk/mechanical-engineering/study/postgraduate-taught/mechanical-engineering-msc

3. "Accreditation of Higher Education Programmes (AHEP)", https://www.engc.org.uk/ahep

4. Pawling, R.J., Bilde, R., Hunt, J., "HYDRA – multipurpose ship designs in engineering and education", International Marine Design Conference (IMDC) 2018, Helsinki, Finland, June 2018.

5. Pawling, R.J., "Studious Ship Design: A Review of UCL MSc Warship Designs", Warship 2019: Multi-Role Vessels, RINA, Bristol, June 2019

6. Brown, D.K. & Moore, G., (2003), "Rebuilding the Royal Navy", London: Chatham Publishing

7. Bradbeer, N. & Manley, D. (2017) "The Application of Wargaming to Education in Naval Design & Survivability" Connections UK, 7 Sep 2017, London

11. THE DESIGN GROUPS

Antunes, M., Martin, R., Morales, R.N. & Thompson, M., "Group 1 – AAW UxV Mothership", Ship Design Project 2020

Peirce, H., Balfour, W., Sng, R., & McKenna, M., "Group 2 – Future LXX", Ship Design Project 2020

Godinho, M.D., Johannessen, H.H., Lacey, N. & Smith, B.V.R., "Group 3 – Depot Ship and Associated Large USV", Ship Design Project 2020

Branch, R., Curtis, S. & Wu, J., "Group 4 – Upgradeable OPV", Ship Design Project 2020

Upton, T., Wilcock, M. & Sinha, R., "Group 1 – Ultra-Stealthy Surface Combatant", Ship Design Project 2021

Dhesi, K., Zhang, J. & McNaught, L., "Group 2 – Persistence Carrier", Ship Design Project 2021

Houlton, S., Makris, D., Wang, S. & Lim, J.H., "Group 3 – Multipurpose Logistics Vessel", Ship Design Project 2021

Martin, L., Mehboob, M. & Fennelly, W., "Group 4 – Optionally Crewed Territorial Defence Vessel", Ship Design Project 2021

Carr, M., Maqsood, B. & McGuire, C., "Group 5 – Common Hull Surface Combatant", Ship Design Project 2021

Dr. Rachel Pawling is a Lecturer in Ship Design at UCL, teaching the subject of ship design to undergraduate and postgraduate students. She obtained her PhD in computer aided ship design in 2007 and has subsequently continued her research in projects funded by the EC, UK and US governments.

Appendix A: Example User Requirements Document for the UCL MSc Ship Design Exercise

OPTIONALLY CREWED TERRITORIAL DEFENCE VESSEL
1. Role: A small, high-speed vessel for patrol and defence of home waters against a numerically superior adversary.

2. Constraints
a) The customer: The defence procurement arm of a technologically advanced nation with a regional security and national defence focus.
b) Emissions and environmental impact: The vessel shall be designed for a future of exceptionally stringent emissions regulations in peacetime and Operations Other Than War (OOTW).
c) Technology insertion: Assume two extended maintenance periods for upkeep, update and upgrade.
d) Survivability: Crew survivability is paramount. The vessel shall retain sufficient FLOAT and MOVE capabilities after a hit from a small ASM to return to port.
e) Signatures: Shall have low above water and should have low below water signatures.
f) Choice of standards: Standards shall be military, which in this case includes LR or DNV-GL naval ship rules.
g) Disposal; Consider scrapping/disposal strategy including environmental impact and recycling.
h) Aesthetics: The customer regards a striking, aggressive appearance as important in illustrating its willingness to defend its waters.
i) Displacement cap: Displacement should not exceed 2000 tonnes standard
j) Stores endurance: Design for 30 days stores endurance.

3. Special Studies
a) Automation: The customer is seeking to exploit automation to radically reduce the core crew required to operate the vessels.
b) The vessel shall be capable of being used in a fully autonomous role augmenting a crewed squadron leader.

4. Primary Tasks
a) Provide fire support to forces ashore using missiles, rockets or guns.
b) Conduct board, inspect, search & seizure operations against compliant and non- compliant merchant vessels including fishing vessels.
c) Anti-ship combat and land attack using heavyweight missiles; defeating an enemy vessel up to destroyer size.

5. Secondary Tasks
a) Undertake defence diplomacy and international engagement by visiting and exercising with friendly and allied nations.
b) Provide naval presence and surveillance, maintaining a surface and air picture.
c) Defend self or escorted vessels against attack by swarming small boats, USVs or UAVs.
d) Defend self against torpedo attack.
e) Deploy minehunting UXVs.
f) Shall achieve a 90%+ chance of successfully defending own ship against attack by 2 simultaneous supersonic anti-ship missiles.

6. Area of Operations
a) Home waters are in Northern Europe. Includes operations in broken ice.
b) Occasional out-of-area deployments will take place, but can assume host nation support.

7. Endurance & Resupply
a) Deployment endurance: maximum of two-month deployments on out-of-area tasks.
b) Fuel endurance: choose an endurance coherent with the mission requirements
c) No requirement to RAS, but ability to rapidly re-store when alongside in civilian "ports of opportunity" is required.
d) Future fuels: Shoreside supply and infrastructure for future fuels may be assumed to be in place.

8. Complement & Accommodation
a) Core complement should be the minimum possible, with no requirement to maintain at action stations for long durations. Alternative and technological approaches to damage control and crew survivability are encouraged.
b) Role-specific mission crew may be taken aboard as required, but should be minimised.
c) Accommodation shall be of a standard typical of a modern western navy, for both core and embarked personnel.

9. Power & Propulsion System
a) A range of power & propulsion options should be considered, and the final choice justified relative to other systems.
b) Electrical Systems: Should consider the choice of AC and DC distribution architectures and make a justified choice. New machine types such as TFM and superconducting motors should be considered for electric propulsion options.
c) Other: Design to take a laser close-in weapon system as an upgrade.

10. Ship Systems
a) Reliability: The ARM philosophy and machinery types should be chosen consistent with a minimum of at-sea maintenance. The ability of the vessel to continue operations with degraded machinery and systems is vital.
b) Maintainability: The use of robotics for inspection, maintenance and damage control shall be considered.
c) Littoral operations: Shall be able to operate in contaminated environment (natural or anthropogenic).

11. Ship Life & In Service Date
a) Ship life: design for a life of 20 years.
b) In service date: 2030

12. Class Size & Budget
a) See costing requirements document.

The Royal Institution of Naval Architects
8-9 Northumberland Street
London, WC2N 5DA
United Kingdom

ISBN 978-1-7138-8309-8

9 781713 883098

CURRAN ASSOCIATES INC.
proceedings.com